Ilaria Menapace

Photoluminescence properties of heat-treated silicon-based polymers

Ilaria Menapace

Photoluminescence properties of heat-treated silicon-based polymers

Promising materials for LED applications

Südwestdeutscher Verlag für Hochschulschriften

Impressum/Imprint (nur für Deutschland/ only for Germany)
Bibliografische Information der Deutschen Nationalbibliothek: Die Deutsche Nationalbibliothek verzeichnet diese Publikation in der Deutschen Nationalbibliografie; detaillierte bibliografische Daten sind im Internet über http://dnb.d-nb.de abrufbar.

Alle in diesem Buch genannten Marken und Produktnamen unterliegen warenzeichen-, markenoder patentrechtlichem Schutz bzw. sind Warenzeichen oder eingetragene Warenzeichen der jeweiligen Inhaber. Die Wiedergabe von Marken, Produktnamen, Gebrauchsnamen, Handelsnamen, Warenbezeichnungen u.s.w. in diesem Werk berechtigt auch ohne besondere Kennzeichnung nicht zu der Annahme, dass solche Namen im Sinne der Warenzeichen- und Markenschutzgesetzgebung als frei zu betrachten wären und daher von jedermann benutzt werden dürften.

Verlag: Südwestdeutscher Verlag für Hochschulschriften Aktiengesellschaft & Co. KG
Dudweiler Landstr. 99, 66123 Saarbrücken, Deutschland
Telefon +49 681 37 20 271-1, Telefax +49 681 37 20 271-0
Email: info@svh-verlag.de
Zugl.: Darmstadt, TU, Diss., 2010

Herstellung in Deutschland:
Schaltungsdienst Lange o.H.G., Berlin
Books on Demand GmbH, Norderstedt
Reha GmbH, Saarbrücken
Amazon Distribution GmbH, Leipzig
ISBN: 978-3-8381-1679-2

Imprint (only for USA, GB)
Bibliographic information published by the Deutsche Nationalbibliothek: The Deutsche Nationalbibliothek lists this publication in the Deutsche Nationalbibliografie; detailed bibliographic data are available in the Internet at http://dnb.d-nb.de.

Any brand names and product names mentioned in this book are subject to trademark, brand or patent protection and are trademarks or registered trademarks of their respective holders. The use of brand names, product names, common names, trade names, product descriptions etc. even without a particular marking in this works is in no way to be construed to mean that such names may be regarded as unrestricted in respect of trademark and brand protection legislation and could thus be used by anyone.

Publisher: Südwestdeutscher Verlag für Hochschulschriften Aktiengesellschaft & Co. KG
Dudweiler Landstr. 99, 66123 Saarbrücken, Germany
Phone +49 681 37 20 271-1, Fax +49 681 37 20 271-0
Email: info@svh-verlag.de

Printed in the U.S.A.
Printed in the U.K. by (see last page)
ISBN: 978-3-8381-1679-2

Copyright © 2010 by the author and Südwestdeutscher Verlag für Hochschulschriften Aktiengesellschaft & Co. KG and licensors
All rights reserved. Saarbrücken 2010

Acknowledgments

The present PhD thesis outlines the highlight of the work that I performed at the TU Darmstadt between September 2006 and December 2009.

Special thanks are for Prof. Ralf Riedel, who offered me the possibility to work in a European project, the "Polycernet", inside the Marie Curie frame, where I could attend many workshops and have collaborations at a European scale. Moreover, during my PhD, I could independently develop my scientific work and participate to international conferences.

I thank Gabriela Mera, who assisted me in my whole PhD, and contributed to the work especially from the chemistry point of view. I also thank Aitana Tamayo, who helped me during the writing of my thesis, and the whole Disperse Feststoffe group, with whom I spent the last three years, my officemates Magdalena and Ricardo, the colleagues Liviu, Christoph, Dmytro, Sandra, Emanuel, Rodrigue, Vassilios, Ravi, Marta, Carmen, Miria, Benjamin, Aleksander, Marina, Mahdi, Andrzej and Marcin, our technician Claudia, the HiWis and our secretaries Su-Chen and Karen.

Moreover, I would like to thank the students that contributed to this thesis work or to other projects that I carried out during my PhD: Erik, Alexander and Steven.

A warm thank to Prof. Heinz von Seggern, Jörg Zimmermann and Graham Appleby from the group Elektronische Materialeigenschaften, to Ute Liepold from Siemens, Prof. Richard Laine from the University of Michigan and Prof. Michl from the University of Boulder for the helpful discussions on photoluminescence mechanisms, to Emre Erdem and Rüdiger Eichel for the EPR analysis, to Ildiko Balog for Raman measurements, to Matthias Adlung for the q.e. measurements, to Christel Gervais, Florence Babonneau and Alberto Pauletti for the solid state NMR analysis, and finally to Cekdar Vakifahmetoglu and Paolo Colombo for the collaboration on SiCN foams.

I also thank the European Commission (Marie Curie Research and Training Network, contract number MRTN-CT-019601, "Polycernet") for financial support.

A particular thank to Marco, my family and my friends.

Table of contents

1. Abstract 1
2. Zusammenfassung 3
3. Introduction and Motivation 5
   3.1. References 10
4. Literature review 12
   4.1. Starting materials 12
      4.1.1. Precursors 12
         4.1.1.1. Polysiloxanes 14
            4.1.1.1.1. Properties and applications 14
            4.1.1.1.2. Synthesis 16
         4.1.1.2. Polysilazanes 17
            4.1.1.2.1. Properties and applications 17
            4.1.1.2.2. Synthesis 17
         4.1.1.3. Polysilylcarbodiimides 18
            4.1.1.3.1. Properties and applications 18
            4.1.1.3.2. Synthesis 19
      4.1.2. Crosslinking reactions 20
      4.1.3. Pyrolysis reactions 22
   4.2. Luminescent silicon-based polymers or polymer derived materials 23
      4.2.1. Luminescent polysilanes and polycarbosilanes 23
      4.2.2. Other luminescent silicon-based polymers or polymer derived materials ... 25
   4.3. References 28
5. Experimental procedure 35
   5. 1. Starting materials 35
   5.2. Heat-treatment 36
   5.3. Photoluminescence measurements 37
      5.3.1. Photoluminescence spectrometry 37
      5.3.2. Lifetime measurements 38

I

5.3.3. Quantum efficiency ... 39
5.4. Absorption measurements ... 39
   5.4.1. UV-Visible spectroscopy ... 39
   5.4.2. Remission/Reflection measurements ... 40
      5.4.2.1. Remission measurements ... 41
      5.4.2.2. Reflection measurements ... 42
5.5. Structural characterizations ... 42
   5.5.1. FT-IR spectroscopy ... 43
   5.5.2. Raman spectroscopy ... 43
   5.5.3. NMR spectroscopy ... 43
      5.5.3.1. Liquid State NMR ... 43
      5.5.3.2. Solid State NMR ... 44
5.6. Thermal analysis ... 44
5.7. X-Ray Diffraction (XRD) ... 45
5.8. EPR spectroscopy ... 45
5.9. Note ... 47
5.10. References ... 47
6. Results and discussion ... 49
   6.1. Polysiloxanes ... 49
      6.1.1. Wacker-Belsil® PMS MK polymethylsilsesquioxane ... 49
         6.1.1.1. Photoluminescence measurements ... 51
            6.1.1.1.1. Fluorescence measurements ... 51
            6.1.1.1.2. Effect of the annealing atmosphere ... 55
            6.1.1.1.3. Stability of the fluorescence during storage ... 57
            6.1.1.1.4. Quantum efficiency ... 58
         6.1.1.2. Absorption measurements ... 59
            6.1.1.2.1. UV-Vis-NIR spectroscopy ... 59
            6.1.1.2.2. Remission measurements ... 60
            6.1.1.2.3. Reflection measurements ... 61
         6.1.1.3. FT-IR spectroscopy ... 62
         6.1.1.4. Raman spectroscopy ... 63
         6.1.1.5. TGA/DTG/MS ... 65
            6.1.1.5.1. MK heat-treated at 500 °C ... 67
         6.1.1.6. XRD measurements ... 68

6.1.1.7. Multinuclear Solid State MAS NMR spectroscopy (^{1}H, ^{13}C and ^{29}Si) 69
    6.1.1.7.1. ^{1}H MAS NMR 71
    6.1.1.7.2. ^{13}C CP MAS NMR 71
    6.1.1.7.3. ^{29}Si MAS NMR 72
    6.1.1.7.4. Discussion of the Solid State MAS NMR results 72
6.1.1.8. EPR measurements 73
6.1.1.9. Discussion 78
6.1.2. Polydimethylsiloxane 85
6.1.3. Polymethylvinylsiloxane 87
6.2. Polysilazanes 90
6.2.1. KiON Ceraset® PUVMS 90
  6.2.1.1. Photoluminescence measurements 92
    6.2.1.1.1. Fluorescence measurements 92
    6.2.1.1.2. Stability of the fluorescence during storage 96
    6.2.1.1.3. Quantum efficiency 97
  6.2.1.2. Absorption measurements 98
    6.2.1.2.1. UV-Vis-NIR spectroscopy 98
    6.2.1.2.2. Remission measurements 99
    6.2.1.2.3. Reflection measurements 100
  6.2.1.3. FT-IR spectroscopy 101
  6.2.1.4. Raman spectroscopy 103
  6.2.1.5. TGA/DTG/MS 106
    6.2.1.5.1. Ceraset heat-treated at 500 °C 107
  6.2.1.6. XRD measurements 108
  6.2.1.7. Multinuclear Solid State MAS NMR spectroscopy (^{1}H, ^{13}C and ^{29}Si) 109
    6.2.1.7.1. ^{1}H MAS NMR 111
    6.2.1.7.2. ^{13}C CP MAS NMR 112
    6.2.1.7.3. ^{29}Si MAS NMR 113
    6.2.1.7.4. Discussion of the Solid State MAS NMR results 113
  6.2.1.8. EPR measurements 114
  6.2.1.9. Discussion 118
6.2.2. KiON VL20 120
  6.2.2.1. Photoluminescence measurements 121
    6.2.2.1.1. Fluorescence measurements 121
    6.2.2.1.2. Stability of the fluorescence during storage 124
    6.2.2.1.3. Quantum efficiency 125
  6.2.2.2. Absorption measurements 126

6.2.2.2.1. UV-Vis-NIR spectroscopy ... 126
6.2.2.2.2. Reflection measurements ... 127
6.2.2.3. FT-IR spectroscopy ... 128
6.2.2.4. Raman spectroscopy ... 130
6.2.2.5. TGA/DTG /MS ... 132
6.2.2.6. XRD measurements ... 134
6.2.2.7. Multinuclear Solid State MAS NMR Spectroscopy (^{1}H, ^{13}C and ^{29}Si) ... 134
    6.2.2.7.1. ^{1}H MAS NMR ... 136
    6.2.2.7.2. ^{13}C CP MAS NMR ... 136
    6.2.2.7.3. ^{29}Si MAS NMR ... 137
    6.2.2.7.4. Discussion of the Solid State MAS NMR results ... 137
6.2.2.8. Discussion ... 138

6.2.3. KiON S ... 139
6.2.3.1. Photoluminescence measurements ... 139
    6.2.3.1.1. Fluorescence measurements ... 139
    6.2.3.1.2. Stability of the fluorescence during storage ... 142
    6.2.3.1.3. Quantum efficiency ... 143
6.2.3.2. Absorption measurements ... 144
    6.2.3.2.1. UV-Vis-NIR spectroscopy ... 144
    6.2.3.2.2. Remission measurements ... 145
    6.2.3.2.3. Reflection measurements ... 146
6.2.3.3. FT-IR spectroscopy ... 147
6.2.3.4. Raman spectroscopy ... 148
6.2.3.5. TGA/DTG /MS ... 149
6.2.3.6. XRD measurements ... 151
6.2.3.7. Liquid state NMR Spectroscopy (^{1}H, ^{13}C and ^{29}Si) ... 152
6.2.3.8. Discussion ... 153

6.3. Polysilylcarbodiimides ... 155
6.3.1. Phenyl-Containing Polysilylcarbodiimides ... 155
6.3.1.1. Photoluminescence measurements ... 156
    6.3.1.1.1. Fluorescence measurements ... 156
6.3.1.2. Absorption measurements ... 164
    6.3.1.2.1. UV-Vis-NIR Spectroscopy ... 164
6.3.1.3. FT-IR spectroscopy ... 166
6.3.1.4. Raman spectroscopy ... 168
6.3.1.5. TGA/DTG/MS ... 169
6.3.1.6. XRD measurements ... 173

6.3.1.7. Discussion 173
6.4. Copolymers 178
  6.4.1. Polydiphenylsilylcarbodiimide + Polysilazane VL20 178
    6.4.1.1. Molecular structure..178
      6.4.1.1.1. FT-IR spectroscopy 178
      6.4.1.1.2. Raman spectroscopy 180
      6.4.1.1.3. XRD measurements 181
      6.4.1.1.4. Liquid State NMR spectroscopy (^{1}H, ^{13}C and ^{29}Si) 182
      6.4.1.1.5. Discussion on the molecular structure 184
    6.4.1.2. Photoluminescence measurements 187
      6.4.1.2.1. Fluorescence measurements 187
      6.4.1.2.2. Stability of the fluorescence during storage..192
      6.4.1.2.3. Quantum efficiency 193
    6.4.1.3. Absorption measurements 194
      6.4.1.3.1. UV-Vis-NIR spectroscopy..194
      6.4.1.3.2. Reflection measurements 196
    6.4.1.4. FT-IR spectroscopy 197
    6.4.1.4. Raman spectroscopy..198
    6.4.1.5. TGA/DTG/MS...199
    6.4.1.6. XRD measurements 202
    6.4.1.7. Multinuclear Solid State MAS NMR spectroscopy (^{1}H, ^{13}C and ^{29}Si) 202
      6.4.1.7.1. ^{1}H MAS NMR 204
      6.4.1.7.2. ^{13}C CP MAS NMR 204
      6.4.1.7.3. ^{29}Si MAS NMR...205
      6.4.1.7.4. Discussion of the Solid State MAS NMR results 205
    6.4.1.8. Discussion 206
6.5. References 210
7. Conclusion 218
8. Outlook 221

1. Abstract

For the development of photoluminescent materials for LEDs, mouldability, insensitivity to air and moisture and resistance to high temperatures are of primary importance, besides improved optical properties. In the present work, a new approach to obtain homogeneous, thermally stable and formable photoluminescent materials starting from silicon-based polymers through heat-treatment at low temperatures (200-700 °C) is presented. These commercially available or laboratory synthesized polymers are usually employed in the PDC route and annealed at temperatures beyond 1000 °C. In the present work, the polysiloxane Wacker-Besil®PMS MK, the polysilazanes KiON Ceraset (PUVMS), KiON VL20 and KiON S, four polysilylcarbodiimides and a copolymer polysilazane-polysilylcarbodiimide were heat-treated up to 700 °C for 2 h under Ar atmosphere. The structural rearrangements during annealing lead to materials with tunable photoluminescence properties.

The maximum emission spectra of many of the polymers investigated show a bathochromic shift as the annealing temperature increases, indicating the development of new luminescent centers with smaller band gap.

For MK, Ceraset and VL20, UV emission is detected after heat-treatment up to 400 °C, and it was observed to be strictly related to the crosslinking extent of the polymers. For heat-treatments from 500 °C, the materials emit in the visible range and this was related to the formation of aromatic agglomerations, precursors of the free carbon phase. At 700 °C for MK and at 600 °C for Ceraset and VL20, the emission range is further red-shifted and the intensity is decreased, due to the concentration quenching related to the development of free carbon in higher amounts.

In the dried KiON S, a visible emission is obtained after annealing at a temperature as low as 300 °C, but the aromatic solvent, still present after drying, has an influence on the fluorescence properties.

All four polysilylcarbodiimides show luminescence properties in the polymer state, due to the presence of phenyl groups. As the treatment temperature increases, red-shifted emission ranges are analyzed for the two polymers which can be crosslinked, while no red-shift was observed for the ones which cannot. After annealing at 500 °C, the fluorescence is quenched in all polymers due to the development of free carbon in high amounts.

1. Abstract

The copolymer polysilazane-polysilylcarbodiimide is very interesting because it shows the maximum emission intensity among all heat-treated polymers analyzed, in the visible range and after heat-treatment at a temperature of just 200 °C. A new luminescent center arises after annealing, formed via crosslinking of the polymer. At 600 °C the luminescence is quenched due to the formation of free carbon.

Therefore, our studies have clearly shown that the luminescence properties are related to structural rearrangements occurring during the polymer to the ceramic transformation, namely crosslinking and formation of sp^2 carbon. The formation of dangling bonds was also considered as a possible source of luminescence, but was demonstrated not to have an influence on the photoluminescent properties.

2. Zusammenfassung

Für den Einsatz von photolumineszenten Materialien in LED Anwendungen sind neben verbesserten optischen Eigenschaften, Formbarkeit, Luft-, Feuchtigkeits- und hohe Temperaturbeständigkeit von Vorteil. In dieser Arbeit wird ein neuer Weg präsentiert, um homogene, temperaturstabile und formbare photolumineszente Materialien aus siliziumbasierten Polymeren durch Tempern bei niedrigen Temperaturen (200-700 °C) zu erhalten. Siliziumhaltige Polymere wie Polysiloxane, -silazane oder –silylcarbodiimide werden üblicherweise als Vorstufe zur Herstellung hochtemperaturstabiler SiCO bzw. SiCN-Keramiken verwendet. Die Polymer-Keramik-Transformation erfolgt in Inertgasatmosphäre bei Temperaturen zwischen 800 und 1400 °C.

Im Rahmen unserer Untersuchungen zur Lumineszenz von Si-Polymeren wurden das Polysiloxan Wacker-Besil®PMS MK, die Polysilazane KiON Ceraset (PUVMS), KiON VL20 und KiON S, vier Polysilylcarbodiimide und ein Hybridsystem, bestehend aus dem Polysilazan VL20 und Polydiphenylsilylcarbodiimid, bis zu 700 °C für 2 h unter Ar Atmosphäre temperiert. Dabei konnte gezeigt werden, dass die mit der Temperaturbehandlungen einhergehenden strukturellen Umordnungen zu Materialien mit einstellbaren photolumineszenten Eigenschaften führen.

Die Emissionsspektren der vernetzbaren Polymere zeigen eine bathochrome Verschiebung mit steigender Auslagerungstemperatur, was auf die Entwicklung neuer lumineszenter Zentren mit kleinerer Bandlücke hinweist. Für das MK-Polymer, für Ceraset und VL20 werden UV Emissionen bis zu Temperaturbehandlungen von 400 °C analysiert. Wie gezeigt werden konnte, steht dieses Verhalten im Zusammenhang mit dem Vernetzungsgrad der Polymere. Nach Tempern bei 500 °C emittieren die Materialien im sichtbaren Bereich, was auf die Bildung aromatischer Zentren während des Temperns zurückzuführen ist. Die sich ausbildenden aromatischen Zentren sind darüber hinaus der Ursprung für die nach Auslagerung bei noch höherer Temperatur nachgewiesene Bildung freier Kohlenstoffphasen. Der Emissionsbereich der höher temperierten Proben (MK bei 700 °C, Ceraset/VL20 bei 600 °C) ist noch weiter rotverschoben, und die Intensität sinkt wegen des Konzentrations-Quenchings aufgrund der zunehmenden Ausscheidung von freiem Kohlenstoff. Im getrockneten KiON S-Polymer wird eine

2. Zusammenfassung

sichtbare Emission schon nach einer Temperierung bei 300 °C erhalten, restliches aromatisches Lösungsmittel beeinflusst jedoch die fluoreszenten Eigenschaften.

Die synthetisierten Polysilylcarbodiimide zeigen auch ohne Temperaturbehandlung aufgrund der Phenylsubstituenten am Silizium lumineszente Eigenschaften im Polymerzustand. Mit steigender Temperierung werden rotverschobene Emissionsbereiche für diejenigen Polymere detektiert, die geeignete Funktionen zu Vernetzung aufweisen. Im Unterschied hierzu wird keine bathochrome Verschiebung für die Polymere festgestellt, die nicht vernetzungsfähig sind. Nach Auslagerung der Polysilylcarbodiimide bei 500 °C vermindert sich die Fluoreszenz aufgrund der zunehmenden Bildung von freiem Kohlenstoff.

Das Hybridsystem Polysilazan/Polysilylcarbodiimid zeigt die höchste Emissionsintensität unter allen ausgelagerten Polymeren und emittiert im sichtbaren Bereich bereits nach einer Temperierung von 200 °C. Auch hier entsteht während der Auslagerung durch Vernetzung des Polymers ein neues lumineszentes Zentrum. Nach einer Temperaturbehandlung bei 600 °C schwächt sich die Lumineszenz wieder ab aufgrund der Ausbildung von freiem Kohlenstoff.

Zusammenfassend kann festgestellt werden, dass die lumineszenten Eigenschaften auf die strukturellen Umordnungen zurückzuführen sind, die bei der Auslagerung der hier untersuchten Polymere im Temperaturbereich zwischen Raumtemperatur und 600 °C erfolgen. Die Bildung von „Dangling Bonds" (ungebundene, freie Valenzen) wurde zunächst als mögliche Ursache für die Lumineszenz der Polymere diskutiert. Eine eindeutige Korrelation der mittels EPR-Spektroskopie ermittelten Spinkonzentration mit der Fluoreszenzintensität konnte jedoch nicht nachgewiesen werden. Aus unseren Untersuchungen geht vielmehr klar hervor, dass die Lumineszenz mit dem Vernetzungsgrad und mit der Bildung von sp^2-Kohlenstoff in den temperaturbehandelten Si-Polymeren korreliert.

3. Introduction and Motivation

The natural luminescence phenomena, such as those shown by fireflies and glow-worms, have fascinated the humans since pre-history, but the oldest documents referring to them date back to 1500-1000 B.C. in China [Harvey1957]. Later on, the development of alchemy in Europe in the 16th and 17th centuries permitted a rational approach toward the study of natural phenomena and the first observation of fluorescence was reported in 1565, referring to the emission of light by an infusion of wood Lignum Nephriticum (by Monardes) [Harvey1957, Valeur2002]. Nevertheless, the most famous example of a luminescent mineral (and first reported observation of phosphorescence) was the "Bolognian Stone" or "Litheophosphorus", discovered in 1602 by the cobbler and dilettante alchemist Vincenzo Casciarolo, during a walk on the Monte Paterno near Bologna [Harvey1957]. This natural stone became the first object of scientific study of luminescent phenomena. After calcination with coal and exposition to the sun, this stone could emit light in the darkness [Harvey1957]. Afterward, it was discovered that the stone contained barium sulfate, which was reduced by the coal to barium sulfide, which is responsible for the phosphorescence. In 1640, the professor of Philosophy at the University of Bologna Fortunius Licetus wrote the text, "Litheosphorus Sive De Lapide Bononiensi", where the first definition as a non-thermal light emission was given:

… ex arte calcinati, et
illuminato aeri seu solis
radiis, seu flammae
fulgoribus expositi, lucem
inde sine calore
concipiunt in sese; …

[… properly calcinated, and
illuminated either by
sunlight or flames, they
conceive light from
themselves without heat; …][Valeur2002].

Licetus,1640

The word luminescence (from the Latin lumen = light) was introduced for the first time by the physicist and science historian Eilhardt Wiedemann in 1888, to describe 'all those phenomena of light which are not solely conditioned by the rise in temperature', as opposed to incandescence [Valeur2002]. Luminescence is then defined as "cold light" in contraposition to "hot light" of incandescence.

Luminescence is an emission of ultraviolet, visible or infrared photons from an electronically excited species. The exciting radiation promotes the activator (molecule or luminescent center) to an electronically excited state, which decays back to the ground state by emission of radiation. If there is a competitor to the radiative emission process, non-radiative return to the ground state could occur: the energy of the excited state is used to excite the vibrations of the host lattice and is lost in heat [Blasse1994].

Luminescence is classified according to the mode of excitation. Photoluminescence (fluorescence, phosphorescence and delayed fluorescence) is related to the absorption of light (photons), radioluminescence to ionizing radiation (X-rays, α, β, γ) (X-ray luminescence to X-rays), cathodoluminescence to cathode rays (electron beams), electroluminescence to electric field, chemiluminescence to chemical processes (e.g. oxidation), bioluminescence to biochemical processes, triboluminescence to frictional and electrostatic forces, sonoluminescence to ultrasounds. Moreover, thermoluminescence is thermal stimulation of luminescence (and not thermal excitation), by means of heating after prior storage of energy (excitation achieved in a different way). The stimulated emission is a transition from the upper to the lower state induced by radiation field [Valeur2002, Blasse1994].

Although the modern scientific approach has produced a vast knowledge of luminescent phenomena, the several papers published every day show that the identification of the origin of luminescence in newly discovered systems is not an easy task. Therefore, luminescence remains an intriguing field of study that still reserves mysteries [Roda1998, Valeur2002].

The present work contributes to the fascinating world of luminescence, presenting a new class of fluorescent materials. Moreover, it investigates a new property of the heat-treated silicon-based polymers, unknown so far, with great potential for application.

The main final application of photoluminescent heat-treated silicon-based polymers is in the field of Light Emitting Diodes (LEDs). In recent years these devices have found applications as display panels, illumination, traffic lights, flashlights, Christmas lights and automotive lights. "Cold light", as opposed to light emitted as thermal radiation of incandescent bulbs, offers the advantage of lower power consumption, higher reliability, increased lifetime and environmental-friendly characteristics [Zhu2007].

3. Introduction and Motivation

Inorganic LEDs are based on a semiconductor chip coated with a light converting material (phosphor), which converts the near-UV or blue light (360-460 nm) into visible light [Rohwer2003].

To improve the conversion LEDs performances, there is constant research for better color quality, higher efficiency and low cost. The research goals of the converting materials are transparency, higher thermal stability and formability.

The most widely used phosphors for LEDs are in the form of a powder suspended in a resin. However, they still encounter problems of sedimentation during the processing [Tan2004, Bert2007]. Moreover, the resin matrix is subjected to photodegradation under UV light and to damage under moisture and at high temperatures [Janet2007, Naredan2004].

Rare earth doped glasses avoid these problems by offering homogeneity, transparency and excellent stability against temperature, radiation and moisture, but they require high processing temperatures for the molding (approximately 1250-1500 °C) [Zhu2007, Sun2007]. Similar materials are sol-gel derived glasses, characterized by considerably reduced processing temperatures, but also by a wet technique, which is not suitable for *in situ* molding.

Promising devices are also organic light emitting diodes (OLEDs), made either of small molecules or π-conjugated polymers. They have undergone improvements in performance and offer an alternative to commercial LEDs, due to their advantages such as ease of production, formability, low operating voltages, low cost, tunability of color emission and flexibility. Nevertheless, they still present problems such as air and moisture sensitivity and instability under high-energy radiation and high temperatures, with consequently short lifetime [Gardonio2007]. The main application field of OLEDs is in flat panel displays [Burrows1996].

Besides the π-conjugated polymers, also the σ-conjugated polymers have been proposed as candidates for light emitting diodes, especially for UV emission. Polysilanes show larger band gaps than organic polymers. They exhibit several advantages such as formability and flexibility, but they are sensitive to air, moisture, high energy radiation and high temperatures [Schubert2004, Suzuki1996, Michl1988, Sharma2006, Sharma2007].

Therefore, the development of light converting materials which are insensitive to air and moisture, thermally stable, homogeneous and easy to process is still a challenging research task. In this thesis, a new approach to obtain thermally stable, homogeneous and formable photoluminescent materials is presented. The materials are obtained by heat-treatment of cheap, non toxic and commercially available polyorganosiloxanes and polyorganosilazanes, synthesized polysilylcarbodiimides and a copolymer based on a polysilazane and a polysilylcarbodiimide.

The thermal stability of Si-based polymers was shown in numerous reports [Schubert2004, Thames1996, Riedel2006]. The presence of silicon in the polymer chains provides heat and radiation resistance, which render these materials more stable than organic polymers and increase their lifetime. Silicones are also water resistant, chemical resistant and oxidation resistant and could be employed in harsh environments [Schubert2004].

Silicon-based polymers are largely used to produce Polymer Derived Ceramics (PDCs). PDCs are a class of ceramic materials created by heat-treating a starting polymer (precursor) at high temperature (1000-2000 °C) [Riedel2006, Li2000a, Li2001, Radovanovic1999, Kleebe2006, Andronenko2006, Laine1995, Trassl2002a, Trassl2002b, Riedel1996a, Riedel1996b, Riedel1992, Berger2005, Berger2004, Jeschke1999, Soraru1990, Richter1997]. The properties of the end ceramic can be controlled by the design and synthesis of the precursor.

Low temperature heat-treatment (200-500 °C) of the ceramic precursors, which are in liquid or powder form, is effective in order to crosslink the inorganic polymer and obtain a transparent and homogeneous solid material which can retain the shape of a mold [Li2000a, Radovanovic1999, Li2001]. The crosslinked organosilicon polymers, after functionalization in order to obtain luminescence properties, could be applied as luminescent material. Disadvantages such as sedimentation, non-homogeneity and high temperature molding can be overcome by using traditional organosilicon polymer processing techniques. Consequently, silicon-based polymers combine the temperature stability of glasses with the excellent formability of organic polymers and may provide attractive alternative materials for the production of light emitting diodes [Suzuki1996, Michl1988]. Additionally, the dry process is an advantage over the wet sol-gel technique and allows *in situ* mouldability. The luminescent properties may be either intrinsic or provided by a suitable dopant. The requirements for a material to be useful in LEDs are stability against temperature (up to 150 °C), UV radiation and moisture, excitability with wavelength in the near-UV or blue range (360-460 nm) and emission in the visible range (400-700 nm) [Rohwer2003]. Transparency is also an advantage. The heat-treated silicon-based polymers could be a solution to these requirements.

Therefore, several silicon polymers were investigated after varying annealing temperatures.

Some of the heat-treated silicon-based polymers analyzed showed unexpected photoluminescence properties. Structural rearrangements during annealing of the Si-based polymers lead to photoluminescence properties in the visible range. To develop, enhance or modify the luminescence properties of the materials, some of the systems were doped with different concentrations of a rare earth element, i.e. europium. Nevertheless, since the first experiments with europium were not successful and the intrinsic fluorescence of undoped materials was by far more attractive, the doping route was abandoned and only the investigation of the intrinsic photoluminescence was performed, also considering the expertise and pursuing the interests of our working group. In some cases, phenyl groups were grafted to the polymer chains. The intense fluorescence emission obtained after heat-treatment of the copolymer based on a polysilazane and a phenyl-containing polysilylcarbodiimide demonstrated the possibility of enhancing the luminescence properties, with great potential for quantum efficiency improvements. Thus, the focus of the research was shifted from the first goal of using the heat-treated silicon polymers as matrix for luminescent dopants to the investigation of the luminescent properties of the annealed silicon polymers, the enhancement of their emission intensity and the clarification of the photoluminescence mechanisms by correlation to the structural properties.

Moreover, the fluorescence analysis provided new information about the processes occurring in the polymers during annealing in the temperature range relative to the transformation polymer-ceramic, useful for the investigation of the PDCs.

Furthermore, the study here presented could be valuable for the development of new polymer-ceramic materials, which have not been properly investigated so far. While the PDCs have found great interest in research [Riedel1992, Riedel2006a, Riedel1996b, Li2000a, Radovanovic1999, Li2001, Kleebe2006, Andronenko2006, Laine1995, Trassl2002a, Trassl2002b, Berger2005], the heat-treated silicon-based polymers, also called ceramers [Wilhelm2005], derived from the same precursors but annealed at lower temperatures, have not been so widely explored. Samples treated at low temperatures were usually inspected in relation to the final ceramic, but were seldom investigated as independent materials [Wilhelm2005, Li2000a, Li2001, Kleebe2006, Laine1995, Berger2005, Berger2004, Soraru1990].

Since the present thesis describes a fundamental work about heat-treated silicon-based polymers and their photoluminescence properties, various polymers were investigated, in order to have an overview on the diverse systems and to explore affinities and differences.

The silicon polymers here considered are a solid commercially available polysiloxane (Wacker-Belsil® PMS MK polymethylsilsesquioxane)[Wacker], two liquid commercial polysilazanes (polyureavinylmethylsilazane Kion Ceraset® PUVMS and KiON VL20 [KiON]), a polysilazane commercially available in solution (KiON S [KiON]), four polyphenylsilylcarbodiimides synthesized in our laboratory [Mera2009b] and a copolymer derived from the reaction of a polysilazane with a polysilylcarbodiimide.

3.1. References

[Andronenko2006] S. I. Andronenko, I. Stiharu, S. K. Misra, *J. Appl. Phys.*, 99, *113907*, 2006
[Berger2004] F. Berger, M. Weinmann, F. Aldinger, K. Müller, *Chem. Mater.*, 16, *919-929*, 2004
[Berger2005] F. Berger, A. Müller, F. Aldinger, K. Müller, *Z. Anorg. Allg. Chem.*, 631, *355-363*, 2005
[Bert2007] B. Bert, *US Patent 20070024173*, 2007
[Blasse1994] G. Blasse, B.C. Grabmaier, *Luminescent Materials*, Springer-Verlag, Berlin Heidelberg, 1994
[Burrows1996] P. E. Burrows, S. R. Forrest, S. P. Sibley, M. E. Thompson, *Appl. Phys. Lett.*, 69 [20], 1996
[Gardonio2007] S. Gardonio, L. Gregoratti, P. Melpignano, L. Aballe, V. Biondo, R. Zamboni, M. Murgia, S. Caria, M. Kiskinova, *Org. Electron.*, 8, *37-43*, 2007
[Harvey1957] E. N. Harvey, *History of Luminescence, The American Philosophical Society*, Philadelphia, 1957
[Janet2007] C. B. Y. Janet, *US Patent 20070247060*, 2007
[Jeschke1999] G. Jeschke, M. Kroschel, M. Jansen, *J. Non-Cryst. Solids,* 260, *216-227*, 1999
[KiON] http://www.kioncorp.com/bulletins.html
[Kleebe2006] H. J. Kleebe, G. Gregori, F. Babonneau, Y. D. Blum, D. B. MacQueen, S. Masse, *Int. J. Mater. Res.*, 97 [6], *699-709*, 2006
[Laine1995] R. M. Laine, F. Babonneau, K. Y. Blowhowiak, R. A. Kennish, J. A. Rahn, G. J. Exarhos, K. J. Waldner, *J. Am. Ceram. Soc.,* 78 [1], *137-145*, 1995
[Li2000a] Y. L. Li, R. Riedel, J. Steiger, H. Von Seggern, *Adv. Eng. Mater.*, 2 [5], *290-293*, 2000
[Li2001] Y. L. Li, E. Kroke, R. Riedel, C. Fasel, C. Gervais, F. Babonneau, *Appl. Organomet. Chem.*, 15, *820-832*, 2001
[Mera2009b] G. Mera, R. Riedel, F. Poli, and K. Müller, *J. Eur. Ceram. Soc.*, 29, *2873-2883*, 2009
[Michl1988] J. Michl, J. W. Downing, T. Karatsu , A. J. McKinley, G. Poggi, G. M. Wallraff, R. Sooriyakumaran, R. D. Miller, *Pure Appl. Chem.*, 60 [7], *959-972*, 1988
[Narendran2004] N. Narendran, Y. Gu, J. P. Freyssinier, H. Yu, L. Deng, *J. Cryst. Growth*, 268, *449-456*, 2004
[Radanovic1999] E. Radovanovic, M. F. Gozzi, M. C. Gonçalves, I. V. P. Yoshida, *J Non-Cryst. Solids*, 248, *37-48*, 1999
[Richter1997] R. Richter, G. Roewer, U. Böhme, K. Busch, F. Babonneau, H. P. Martin, E. Müller, *Appl. Organomet. Chem.*, 11, *71–106*, 1997

3. Introduction and Motivation

[Riedel1992] R. Riedel, G. Passing, H. Schönfelder, R. J. Brook, *Nature*, 355, *714-717*, 1992

[Riedel1996a] R. Riedel, A. Kienzle, W. Dressler, L. Ruwisch, J. Bill, F. Aldinger, *Nature*, 382, *796-798*, 1996

[Riedel1996b] R. Riedel, *Processing of Ceramics, Part II, Vol. 17B*, ed. R. J. Brook. VCH, Würzburg, *pp. 1–50*, 1996

[Riedel2006] R. Riedel, G. Mera, R. Hauser, A. Klonczynski, *J. Ceram. Soc. Jpn.*, 114, *425-444*, 2006

[Roda1998] A. Roda, M. Pazzagli, L. J. Kricka, P. E. Stanley, *Bioluminescence & Chemiluminescence: Perspectives for the 21st Century*, John Wiley & Sons Ltd, Chichester, 1998

[Rohwer2003] L. S. Rohwer, A. M. Srivastava, *Electrochem Soc. Inteface*, 12 [2], *36-39*, 2003

[Schubert2004] U. Schubert, N. Hüsing, *Synthesis of Inorganic Materials*, Wiley-VCH, Weinheim, Ch. 5, 2004

[Sharma2006] Sharma, Katiyar, Deepak, Seki, Tagawa, *Appl. Phys. Lett.*, 88, *143511*, 2006

[Sharma2007] Sharma, Katiyar, Deepak, Seki, *Appl. Phys.*, 102, *084506*, 2007

[Soraru1990] G. D. Sorarù, F. Babonneau, J. D. Mackenzie, *J. Mater. Sci.*, 25, *3886-3893*, 1990

[Sun2007] X. Sun, J. Zhang, X. Zhang, S. Lu, X. Wang, *J. Lumin.*, 122-123, *955-957*, 2007

[Suzuki1996] H. Suzuki, *Adv. Mater.*, 8 [8], *657-659*, 1996

[Tan2004] K. L. Tan, A. K. Aizar, S. L. Oon , B. C. Tan, *US Patent 6806658*, 2004

[Thames1996] S. F. Thames, K. G. Panjnani, *J Inorg. Organomet. P.*, 6 [2], *59-94*, 1996

[Trassl2002a] S. Trassl, G. Motz, E. Rössler, G. Ziegler, *J. Am. Ceram. Soc.*, 85 [1], *239-244*, 2002

[Trassl2002b] S. Trassl, H.-J. Kleebe, H. Störmer, G. Motz, E. Rössler, G. Ziegler. *J. Am. Ceram. Soc.*, 85 [5], *1268-1274*, 2002

[Valeur2002] B. Valeur, *Molecular Fluorescence Principles and Applications*, Wiley-VCH, Weinheim, 2002

[Wacker] http://www.wacker.com/cms/de/home/index.jsp

[Wilhelm2005] M. Wilhelm, C. Soltmann, D. Koch, G. Grathwohl, *J. Eur. Cer. Soc.* 25, *271–276*, 2005

[Zhu2007] C. Zhu, Y. Yang, X. Liang, S. Yuan, G. Chen, *J. Lumin.* 126 [2], *707-710*, 2007

4. Literature review

Due to the novel character of the present research topic, no literature review can be presented on the previous state of the art. An introductory literature review can only be performed on the materials used and on luminescent materials similar to the ones investigated here. In the next section, the properties and synthesis methods to produce polysiloxanes, polysilazanes and polysilylcarbodiimides are shown, complete with the description of the crosslinking and pyrolysis reactions that convert the polymers to the ceramic state during heat-treatment. The following section will give an overview about luminescent silicon-based polymers and polymer derived materials.

4.1. Starting materials

4.1.1. Precursors

Inorganic polymers are differentiated from organic polymers on the basis of their backbone, which contains elements different from carbon, nitrogen and oxygen, and to which organic or organometallic substituents are attached [Schubert2004]. The most common classes of inorganic polymers are polysiloxanes, polysilazanes, polysilanes, polycarbosilanes and polyphosphazenes; polysilylcarbodiimides are acquiring increasing importance (Figure 1).

5. Experimental procedure

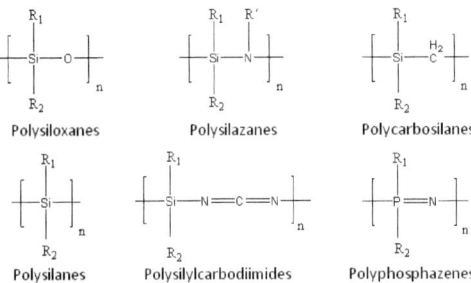

Figure 1: Molecular structural units of some inorganic polymers.

Inorganic polymers combine the advantages of carbon-based polymers and inorganic materials. Organic polymers are light, tough and easy to fabricate, but in general they cannot be exposed to high temperatures, strongly oxidizing conditions, or high-energy radiation, for which inorganic materials are suitable [Schubert2004]. Inorganic polymers are characterized by an inorganic backbone and organic substituents. The backbone can provide heat, fire, or radiation resistance, as well as electrical conductivity or flexibility. The substituents possess the ability to modify properties such as solubility, liquid crystallinity, optical properties, hydrophobicity or hydrophilicity, adhesion or biological compatibility. They also provide the possibility of crosslinking between the polymer chains. Inorganic polymers are mainly used as preceramic materials, because of the advantages of the polymer processing (shaping), in applications as ceramic fibers, films, matrices in composite materials and porous materials [Schubert2004]. The importance of the PDC technique resides in the possibility of tailoring the final ceramic structure by designing the chemical composition of the preceramic polymer. Since the focus of the present thesis is not the synthesis of the preceramic polymers, but their use in the development of new luminescent annealed polymer systems, only a brief review is presented, about the synthesis techniques, the properties and the applications of the silicon-based polymers applied.

The monomers used in the synthesis of organosilicon polymers contain active sites such as Si-Cl, Si-H, Si-C=C and Si-C≡C, which allow polymerization by means of elimination, substitution or addition reactions. The chlorosilanes R_xSiCl_{4-x} (X=0, 1, 2, 3) are the most frequently used starting compounds for the synthesis of polysilanes, polycarbosilanes, polysiloxanes, polysilazanes, polyborosilazanes, polysilylcarbodiimides, polysilsesquioxanes, polycarbosiloxanes etc., due to their commercial availability, low cost and well-known chemistry [Mera2009a] and they are produced by the Müller-

Rochow synthesis on a large scale [Kroke2000]. The natural source of silicon is silica (SiO_2). Firstly, silica is reduced to elemental silicon by the carbothermal method and then transformed into organosilicon species with one of the following routes:
- Reaction of an organic compound with silicon. Silicone industry is mostly based on dimethyldichlorosilane (DDS), which is obtained in the Direct Process or Müller-Rochow process, involving the reaction of gaseous methyl chloride with silicon, containing copper as catalyst in a fluidized-bed reactor at 250-300 °C (Si + CH_3Cl → $(CH_3)_xSiCl_{(4-x)}$). With an optimized process Me_2SiCl_2 is mainly formed, followed by smaller amounts of $MeSiCl_3$, Me_3SiCl and $Me(H)SiCl_2$
- Chlorination of silicon and subsequent substitution of some chlorine atoms by organic groups with organometallic reagents
- Transformation of silicon into silyl hydrides and subsequent addition to multiple bonds in a hydrosilylation process [Schubert2004].

4.1.1.1. Polysiloxanes

4.1.1.1.1. Properties and applications

The polysiloxanes, polymers with Si-O bonds in the main chain, are the oldest and most technologically important class of silicon-based polymers. Since the synthesis of the first polysiloxane carried out by Ladenburg in 1872 [Landenburg1872], most research in silicon-based polymers has been directed towards the synthesis of new polysiloxanes [Zhou2008], the understanding of their properties and the extension of their applications [Chojnowski2000].

Polysiloxanes are largely used as precursors to produce SiOC ceramics, through pyrolysis in inert atmosphere [Riedel2006, Saha2007, Saha2006, Trimmel2003, Gumula2004, Fukushima2004, Lee1999, Latournerie2006, Kleebe2006, Zhang2004, Soraru1994, Brequel2004]. The applications of SiOC ceramics vary from the field of flexible processing of ceramic matrix composites [Harsche2004a, Harsche2004b, Riedel2006] to ceramic micro-electro-mechanical-systems (MEMS) [Harsche2004a, Riedel2006], foams [Colombo1999a, Colombo1999b, Colombo2001, Colombo2004], precision components [Riedel2006], high-temperature resistant coatings for glass and oxide fibers and glow plugs [Riedel2006].

The polysiloxanes are widely employed also in the polymer state, as oils, elastomers (rubbers) and resins [Schubert2004]. Their properties include stability to air, temperature, UV irradiation and weathering, electrical insulation and chemical inertness, hydrophoby, antistick against oil, fats and organic polymers, and high spreadability, good oxygen permeability (valuable in medical applications) and flexibility even at low temperatures. Thanks to the mentioned properties, several technological applications of silicones were developed [Schubert2004, Chojnowski2000].

Silicon oils are used as capacitors and transformers fluids, hydraulic oils, compressible fluids for liquid springs, lubricants, heat transfer media in heating baths, sun-tan lotion, lipstick, shampoo, in cosmetics, as antifoaming and hydrophobizing agents, in cooking oils and in the processing of fruit juices [Schubert2004].

Silicone elastomers are cured silicones, and can be divided into four classes. The first class is characterized by adhesive properties and it is mainly used as adhesives or coating in cars, buildings, industrial plants, household items and in mechanical and electrical engineering; the second type of silicones is characterized by no adhesion to almost any material and applied for sealing and encapsulating electronic components, for vibration absorption, plugs, rubber stamps, for accurate molds and for rapid and flexible impressions, and as foams for fire protection and insulation of cable passages. These two types of silicones are room-temperature vulcanizing (RTV) silicones. The third category of silicones is the liquid rubber, developed for the production of small elastic objects by injection molding, like O-rings, plugs, membranes, gaskets, keyboard switches and the fourth class consists in high-temperature vulcanizing formulations applied as transparent tubing in food industry and medicine, electrical insulations, gaskets, keyboards, oxygen masks [Schubert2004].

Silicone resins are used for electrical insulation, for the encapsulation of components such as resistor and integrated circuits, high-temperature binders in paints and coating for industrial plants, cooking ware, ovens, etc [Schubert2004].

4.1.1.1.2. Synthesis

The polysiloxanes are obtained through the reaction of functional silanes with general formula $R_{4-n}SiX_n$ (n=1, 2, 3), where X is frequently Cl, but it could also be -OR, -OC(O)R, -NR$_2$ or other hydrolyzable groups [Chojnowski2000, Schubert2004].
The route from the halosilane monomers to linear polysiloxanes consists in hydrolysis and condensation of the silane precursor, which lead to a mixture of linear and cyclic oligosiloxanes.
The following reversible reactions represent the hydrolytic polycondensation:

$$\equiv SiCl + H_2O \leftrightarrow \equiv SiOH + HCl$$
$$\equiv SiCl + SiOH \leftrightarrow \equiv SiOSi\equiv + HCl$$

equivalent to

$$2\equiv SiCl + H_2O \leftrightarrow \equiv SiOSi\equiv + 2HCl.$$

Moreover, the homofunctional condensation should also be considered:

$$\equiv SiOH + \equiv SiOH \leftrightarrow \equiv SiOSi\equiv + H_2O.$$

Hydrolytic polycondensation carried out in excess of water occurs via homofunctional silanol condensation, while the heterofunctional polycondensation ($\equiv SiCl + SiOH \leftrightarrow \equiv SiOSi\equiv + HCl$) dominates in the hydrolytic polycondensation of DDS carried out with concentrated HCl solution [Schubert2004].
In another route, methanol substitutes water and the chlorine from the Si-Cl groups is converted into methyl chloride (methanolysis):

$$\equiv SiCl + MeOH \leftrightarrow \equiv SiOH + MeCl.$$

In the presence of hydrolyzable groups such as -OR, the following condensation reaction takes place:

$$\equiv Si\text{-}OH + Si\text{-}OR \equiv \leftrightarrow \equiv Si\text{-}O\text{-}Si\equiv + ROH.$$

The oligomers obtained from the hydrolytic polycondensation are subsequently transformed into a high molecular weight polymer by polycondensation of the hydroxy-ended siloxanes or polysiloxanes or by ring opening polymerization of the cyclic oligomers.
Branched polysilsesquioxanes -[RSi-O$_{1.5}$]$_n$- can also be obtained: they are used as preceramic polymers and have functional properties [Mera2009a].

4.1.1.2. Polysilazanes

4.1.1.2.1. Properties and applications

Polysilazanes are polymers analogous to polysiloxanes, where the oxygen is replaced by NH or NR groups, therefore having Si-N bonds in the main chain.
The first polysilazanes obtained were air and moisture sensitive and, because of the lack of suitable preparative methods for producing linear chains of high molar mass, the polyorganosilazanes have received far less attention than the isoelectronic polysiloxanes [Soum2000]. Nevertheless, they are thermally more stable than polysiloxanes [Schubert2004].
Only in the last years the research activity in the field of polysilazanes has increased, partly thanks to their employment as precursors for silicon nitride and silicon carbonitride (SiCN) through the pyrolysis process. The pyrolysis atmosphere plays a crucial role on the final composition of the materials [Liemersdorf2008]. High pressure nitrogen and ammonia are nitridating reagents, ammonia being more reactive and effective. The SiCN polymer derived ceramics are potential materials for application as wear and corrosion resistant films as well as micro- and optoelectronic devices, porous membranes for hot gas separation (H_2, He) (permselective membrane), foams [Vakifahmetoglu2009, Wang2005], fibers as reinforcement for ceramic materials [Riedel2006] and ceramic micro-igniter (SiOCN) [Riedel2006].

4.1.1.2.2. Synthesis

The polysilazanes are obtained by the reaction of ammonia or amines (ammonolysis or aminolysis) with halogenosilanes. Although the ammonolysis of $SiCl_4$ is useful for the production of Si_3N_4 powders, it is not suitable to obtain fusible or soluble preceramic polymers. Oligomeric or polymeric silazanes are obtained by the reaction of dihalogenosilanes R_2SiCl_2 with ammonia or amines (R_2SiCl_2 + $3R'NH_2$ → -$(SiR_2\text{-}NR')_n$- + $2(R'NH_3)Cl$). The amount of cyclic products is directly proportional to the steric bulk of the substituents both at silicon and nitrogen [Schubert2004, Kroke2000].

During the ammonolysis and aminolysis of chlorosilanes, a large amount of ammonium chlorides is produced. They are difficult to remove from the preceramic polymers and they can act as catalyst for the splitting of the Si-N bonds. To avoid these inconveniences, hexamethyldisilazane is used instead of the amines, and trimethylchlorosilane is obtained as by-product through the Si-Cl/Si-N metathesis reaction (R_2SiCl_2 + Me_3Si-NH-$SiMe_3$ → -(SiR_2-NH)$_n$- + $2Me_3SiCl$) [Schubert2004, Birot1995].

The N-H/Si-H dehydrocondensation, i.e. the formation of Si-N bonds by hydrogen elimination from Si-H and N-H groups, is a convenient method for polymerizing oligomeric silazanes possessing =SiH-NH- groups and it is applied for the most important preceramic polymers.

Ring Opening Polymerization (ROP) of cyclosilazanes leads to high molecular weight linear polysilazanes. The reactivity of the ROP process decreases with the steric bulk of the substituents [Schubert2004].

Transamination reactions are very important in the formation of the N(Si≡)$_3$ nodes during pyrolysis of the polysilazanes and can be employed in the preparation of preceramic polymers without the use of chlorosilanes [Schubert2004]. Trisdimethylaminosilane (HSi(NMe$_2$)$_3$), obtained from elemental silicon and dimethylamine under Müller-Rochow conditions, undergoes transamination reactions in the presence of ammonia or primary amines and a catalyst. The intermediate aminosilanes with Si-NHR and Si-NH$_2$ groups undergo condensation to give Si-NR-Si or Si-NH-Si bonds ((Me_2N)$_3$SiH + $2RNH_2$ → -(HSi(NHR)-NR)$_n$- + $3Me_2NH$) [Schubert2004].

Highly branched silsesquiazane polymers -[RSi-(NH)$_{1.5}$]$_n$- are obtained using trichlorosilanes in the synthesis [Mera2009a].

4.1.1.3. Polysilylcarbodiimides

4.1.1.3.1. Properties and applications

Since the work of Ebsworth, Wannagat and Birkofer on the synthesis of silylcarbodiimide [Ebsworth1961, Ebsworth1962, Pump1962a, Pump1962b, Pump1963, Birkofer1962], several monomeric and polymeric silylcarbodiimides have been reported [Riedel2006, Riedel1998]. Organosilylcarbodiimides have found applications as stabilizing agents for polyurethanes and polyvinylchloride, as insulator coatings, high temperature stable pigments [Klebe1966] and as

irradiation-resistant sealing materials [Razuvaev1987, Riedel2006, Riedel1998]. Moreover, the polysilylcarbodiimides have been used for the synthesis of organic cyanamides, carbodiimides and heterocycles [Gorbatenko1977, Riedel2006, Riedel1998].
Polysilylcarbodiimides are generally air and moisture sensitive materials. By insertion of bulky aromatic substituents at silicon, the air sensitivity massively decreases [Mera2009b]. The phenyl-containing polysilylcarbodiimides were reported to be suitable precursors for high-temperature stable carbon-rich silicon carbonitride polymer-derived ceramics [Mera2009b, Morcos2008a, Riedel2006].

4.1.1.3.2. Synthesis

The first synthesis of polysilylcarbodiimides reported was performed by Pump and Rochow, based on the reaction of diorganochlorosilanes with disilvercyanamide [Riedel1998]. Polysilylcarbodiimides are usually obtained by the reaction of bistrimethylsilylcarbodiimide $Me_3Si-N=C=N-SiMe_3$ and $SiCl_4$ or other cholosilanes in organic solvents with pyridine as a catalyst (by Klebe and Murray). During the reaction Me_3SiCl is eliminated and polymeric $-(Si(N=C=N)_2)_n-$ is formed [Riedel2006, Riedel1998]. Depending on the chlorosilanes used, also highly branched polysilsesquicarbodiimides can be obtained [Mera2009a]. For example, polymethylsilsesquicarbodiimide was produced by the reaction of methyltrichlorosilane with bistrimethylsilylcarbodiimide.
The reaction sequence for the production of silylcarbodiimide polymers is analogous to the hydrolysis and condensation reaction in polysiloxanes, where bistrimethylsilylcarbodiimide acts as the water. The hydrolysis step, when the hydroxyl groups substitute the chlorine bonded to silicon, is analogous to the substitution of chlorine by the silylcarbodiimide unit. Through polycondensation the non-oxidic carbodiimide-containing polymer is obtained [Riedel2006, Riedel1998]. The bistrimethylsilylcarbodiimide is produced by the reaction of cyanamide with chlorotrimethylsilane ($2Me_3SiCl + H_2N-CN + 2Py \rightarrow Me_3Si-NCN-SiMe_3 + 2HCl + 2Py$) or hexamethyldisilazane ($H_2N-CN + (Me_3Si)_2NH \rightarrow Me_3Si-NCN-SiMe_3 + NH_3$) or dicyandiamide (cyanoguanidine, dimer of cynamide) with hexamethyldisilazane ($2(Me_3Si)_2NH + H_2NC(NH)NHCN \rightarrow 2Me_3Si-NCN-SiMe_3 + 2NH_3$) [Riedel1998].

Besides the binary phases SiC and Si_3N_4 generated by the pyrolysis of polysilylcarbodiimides, silicon dicarbodiimide $Si(NCN)_2$ (or SiC_2N_4), produced by the reaction of $SiCl_4$ with bistrimethylsilylcarbodiimide, represents the first crystalline ternary Si-C-N phases. At temperatures higher than 950 °C it decomposes to give the second crystalline SiCN compound known so far, Si_2CN_4 [Mera2009a, Riedel2006, Riedel1998].

4.1.2. Crosslinking reactions

After the synthesis, the polymers are mainly linear, cyclic or weakly branched and often they need to be crosslinked (vulcanized, cured) in order to create bonds between the chains. Crosslinking provides the polymer with higher tensile strength and elastic properties in comparison to a polymer where the chains are free and slide past each other. Heavily crosslinked polymers are rigid, inflexible and infusible. Dissolution of soluble polymers is prevented with a light crosslinking, although the polymer may swell, while with increasing number of crosslinks the degree of swelling decreases [Schubert2004].

Crosslinking is an important step also in the PDC route, because it converts the polymers into infusible materials which retain their shape in pyrolysis. Through the crosslinking stage, the thermoplastic material (polymer), which is meltable, is converted into a thermoset (green body), which is not able to melt anymore [Mera2009a]. The crosslinking is carried out at temperatures up to 400 °C, it improves the thermal stability of the polymers and increases the ceramic yield due to the inhibition of the evolution of volatile products of low molecular mass during pyrolysis [Ionescu2009a, Schubert2004]. Moreover, the polymers can be shaped through simple polymer processing and the production of ceramics pieces with a defined geometry is allowed. Therefore, the pendant groups are important for the modification of the polymers and for the crosslinking process [Mera2009a].

Crosslinking can be performed either thermally or through UV irradiation, using different UV crosslinking agents. Crosslinking can be promoted by addition, condensation, polymerization, metal coordination, radiation and substitution. The addition reaction is the most widely used, mainly carried out with reactive element-hydrogen bonds (for example Si-H) and unsaturated groups (for example Si-CH=CH$_2$) (hydrosilylation). The condensation reactions involve the formation of Si-O-Si or Si-NR-Si bonds starting from Si-OH or Si-NHR groups by elimination of water or ammonia.

5. Experimental procedure

Polymerization or polyaddition reactions occur through unsaturated organic groups in the side chain, for example vinyl groups. Radiation-induced crosslinking is carried out with UV, X-ray, γ or electron radiations, which cause C-H or C-C homolytic cleavage with formation of carbon radicals, which subsequently recombine generating covalent bonds. In this case no functional side groups are required. The substitution reaction is not common, because polymers with displaceable substituents are not readily prepared [Schubert2004].

In polysiloxanes, crosslinking occurs via condensation, free radical initiation reactions or transition metal catalyzed addition. If hydroxy (Si-OH) or alkoxy (Si-OR) groups are present, condensation of silanol groups give rise to Si-O-Si bonds and water, which subsequently hydrolyses the alkoxy substituents. If vinyl groups are present it is possible to thermally crosslink the polymer using peroxides. Furthermore, metal salt can be used for the reaction of Si-H units with vinyl groups (hydrosilylation) [Ionescu2009a].

Polysilazanes obtained by ammonolysis and aminolysis generally possess low molecular weight and crosslinking is needed to avoid the evaporation of oligomers during pyrolysis. They can be crosslinked by heat-treatment or using chemical reagents (catalysts, peroxides). During thermal crosslinking of polysilazanes four reaction steps can take place. In the presence of Si-H and vinyl groups, hydrosilylation reactions occur (at 100-120 °C) and lead to the formation of Si-C-Si and Si-C-C-Si bonds for α or β position respectively [Ionescu2009a, Kroke2000]. Starting from 200 °C, transamination reactions occur, with evolution of amines, ammonia or silazanes and the nitrogen content of the polymer decreases. Dehydrogenation (dehydrocoupling) of Si-H/N-H or Si-H/Si-H groups starts at about 300 °C and leads to Si-N and Si-Si bond formation, with hydrogen evolution. Below 350 °C vinyl polymerization takes place and form carbon chains, which at higher temperatures transform into sp^2 carbon [Kroke2000]. The last crosslinking reaction occurs without mass loss. Crosslinking is increasingly efficient starting from dehydrocoupling of Si-H bonds, vinyl polymerization, transamination, Si-H/N-H dehydrocoupling and finally hydrosilylation [Schubert2004, Birot1995].

The crosslinking of polysilylcarbodiimides depends exclusively on the substituents, as the N=C=N unit is stable up to 500 °C. Nevertheless, the processes have not been yet clarified.

4.1.3. Pyrolysis reactions

During pyrolysis, mineralization and ceramization reactions take place in the temperature range 600-2000 °C. In this process, the crosslinked polymers undergo a transformation into ceramics, by thermolysis and elimination of the organic groups.

The pyrolysis reactions are not yet completely understood, in part because the structure of the preceramic materials, intermediary products and ceramics are also not entirely known. Mineralization occurs at about 500 °C, in correspondence with the first loss of hydrogen and methane, due to methyl-hydrogen and methyl-methyl reactions, and indicates the evolution of organic groups. Ceramization occurs at higher temperatures, at about 800 °C, where hydrogen and methane are further eliminated and rearrangements take place in the ceramic material [Bernard2006, Radovanovic1999].

The polysiloxanes pyrolyzed at temperatures higher than 1000 °C result in SiOC glasses. Between 400 and 600 °C, evolution of CH_4 and hydrogen and redistribution reactions between Si-O, Si-C and Si-H occur, with evolution of low molecular weight silanes. Concurrently with methane and hydrogen elimination free carbon develop, due to C-H and Si-C cleavage and formation of Si• and C• radicals, which recombine and generate Si-C-Si bridges and free carbon. At higher temperatures (600-1000 °C), cleavage of C-H, Si-C and Si-O bonds continues and amorphous silicon oxycarbide and free carbon are formed [Ionescu2009a].

The pyrolysis of polysilazanes generates SiCN ceramics. Also in this case, if hydrogen and methyl groups are substituents on the silicon, starting from 550 °C the reaction between Si-H and Si-CH_3 groups form Si-CH_2-Si units, with methane evolution. At the same time, N-H groups react to form SiN_4 units by replacement of methyl groups and methane elimination. With increasing pyrolysis temperature, the amount of Si-N or Si-C groups increases. When vinyl groups are present, they react at low temperatures (below 350 °C) forming carbon chains which at higher temperatures convert into sp^2 carbon. Reactions between Si-H and Si-CH_3 with N-H groups are responsible for the increasing amount of Si-N bonds with temperature. In the case of the Ceraset, at lower temperatures vinyl polymerization and hydrosilylation occur, while transamination reactions continue up to 600 °C, with ammonia evolution. At higher temperatures (600 to 800 °C) a decrease of Si-H, Si-CH_3 and N-H bonds is observed, with evolution of hydrogen (Si-H/N-H reactions) and methane (Si-CH_3/N-H reactions). The

SiCN ceramics obtained via pyrolysis of polysilazanes consist of a single amorphous silicon carbonitride phase with mixed bonds plus free carbon [Li2001, Ionescu2009a].
On the contrary, after pyrolysis of the polysilylcarbodiimides, SiCN ceramics composed of an amorphous system of silicon nitride and carbon phases without mixed bonds are obtained (as shown by solid state NMR analysis), having a different pyrolytic behavior in respect to polysilazanes. Between 600 and 1000 °C, decomposition of the N=C=N units of the polysilylcarbodiimides occurs, while amorphous Si_3N_4 phase and graphene-like domains start to form [Ionescu2009a, Mera2009b].

4.2. Luminescent silicon-based polymers or polymer derived materials

4.2.1. Luminescent polysilanes and polycarbosilanes

Thanks to their delocalized σ-conjugated electrons along the backbone chain, the polysilanes (polysilylenes) are interesting materials for optoelectronic, for example as light emitting diodes [Suzuki2000, Hua1997]. The σ-conjugation is caused by the overlapping of the Si $3sp^3$ orbitals along the Si-Si chain. Due to their electron state, polysilanes were found to show luminescence properties, mainly in the UV range, correlated to the σ-σ* transition [Sacarescu2008, Ma2007, Sharma2006, Sharma2007, Michl1988, Sun1991]. It was observed that, as the number of silicon atoms in the chain increases, the σ/σ* energy gap becomes smaller, and thus a red-shift in the σ-σ* transition is detected [Schubert2004]. In most polysilanes, aromatic substituents were grafted to the silicon atoms of the chain [Todesco1986, Nespurek2002, Suzuki1998, Suzuki1996, Skryshevskii2003, Wallraff1992]. If aryl substituents are bonded to silicon atoms, the π-orbitals interact with the orbitals of the Si-Si backbone causing an ulterior red-shift [Schubert2004].

Polysilanes were also thermally annealed, with the result of changes in the photoluminescence properties. With increasing heat-treatment temperature and loss of organic groups, bathochromic shift in the photoluminescence emission is observed, until the emission of porous silicon is reached [Singh2008, Watanabe2003]. For the branched polysilanes, with increasing branching structure, a continuous increase in the σ-conjugation and a continuous red-shift of the emission spectrum is expected, but a dual emission was detected. The first emission was attributed to the linear and the second to the branched parts of the Si-Si chain, the latter caused by the formation of a localized state induced by the distortion of the Si-Si chain around the branching point in the excited state. With increasing branching points, the σ-conjugation spreads in two-and three-dimensional directions along the branched Si-Si chain [Watanabe2003, Watanabe2001].

Polysilanes were object of several theoretical studies [Fogarty2003, Rooklin2003]. Besides the application as light emitting diodes, they have the potential for application as thermal precursors to silicon carbide, photo initiators for polymerization reactions (they are photochemically degraded), as photoresists in microlithography, non-linear optics, conductive and semiconductive polymers [Ma2007, Schubert2004].

The relevance of polysilanes resides in the σ-conjugated structure, which is missing in the other silicon-based polymers. Nevertheless, as consequence of the σ-delocalized electronic structure, the polysilanes are easily oxidized and the polymer oxidation is irreversible, causing degradation of the polysilane [Wallraff1992]. Moreover, the polysilanes suffer photochemical degradation when irradiated [Schubert2004, Wallraff1992, Michl1988]. Thus, the instability to air and moisture of polysilanes is a disadvantage [Wallraff1992].

The polycarbosilanes are another class of silicon polymers similar to polysilanes, characterized by the presence of carbon in the polysilane chain. Also the polycarbosilanes are used in the PDC route to produce ceramics but they are also characterized by functional properties if the polymer backbone contains silicon alternated to aromatic groups. In this case the luminescence properties are provided by the π-conjugated system tuned by the presence of the organosilicon unit [Kim1997, Sanchez2008, Rathore2009]. Polycarbosilanes presenting an alternating arrangement of a π-conjugated unit and silicon atoms find applications also as deep-UV photoresists because of their semiconducting properties [Mera2009a].

4.2.2. Other luminescent silicon-based polymers or polymer derived materials

Unlike the polysilanes, the polysiloxanes, polysilazanes and polysilylcarbodiimides are not characterized by σ-conjugated structure. No reports were found about luminescent polysilazanes, although they were applied in electroluminescent devices as hole-transporting mediums [Perry1990], nor about luminescent polysilylcarbodiimides.

Luminescent polysiloxanes have been reported in the literature, provided that a chromophore (anthracene, phenyl) was grafted to the polymeric chain [Suzuki1997, Salom1987a, Salom1987b, Dias2000, Dias2002, Itoh2001, Maçanita1994a, Maçanita1994b, Maçanita1994c, Itoh1996, Maçanita1991, Salom1989, Itoh2002, Vinod2003, Szadkowska-Nicze2004, Toulokhonova2003, Hennecke1988, Shaw1983, Belfield1998, Kim2005, Bisberg1995, Nagai1996, Hamanishi1993]. Correspondingly, pure polysiloxanes, without the addition of luminescent groups, are assumed not to show emissive properties. The polysiloxanes are characterized by thermal, oxidative and chemical stability and, by grafting with aromatic moieties, luminescence properties can also be provided.

Pivin *et al.* reported in 1998 [Pivin1998] and in 2000 [Pivin2000] the fluorescence properties of ion irradiated and/or annealed (1000 °C) polysiloxanes and polycarbosilanes without the insertion of chromophores. It was stated that, in the ceramic thin films characterized, the formation of carbon clusters were responsible for the photoluminescence properties, which vanish with the growth of the carbon segregation [Pivin1998, Pivin2000]. Similar results were found in the present thesis for heat-treated silicon-based polymers and the photoluminescence mechanism attributed to the cited materials is partially in agreement with our findings.

Silicon carbonitride ceramics obtained via the PDC route or by sputtering were also reported to be luminescent [Ferraioli2008, Du2006].

Particular attention has been paid in the literature to luminescent materials from sol-gel reactions, which are suitable for several application areas as waveguides and non-linear optical materials [Gvishi1997]. Plenty of literature exists on luminescent xerogels and aerogels starting from organosilanes [Bekiari1998, Bekiari2001, Bekiari2000, Brankova2003, Stathatos2000, Stathatos2002, Das2007, Das2008]. Also sol-gel derived glasses and phosphors obtained through temperature treatment are known luminescent materials. The heat-treatments vary from simple drying at 50 °C to pyrolysis at 1500 °C [Hreniak2002, Suzuki2006, You2006, Hayakawa2005, Das2007, Das2008, Baran2004]. Obviously, the benefit of the sol-gel technique is the low temperatures required compared to the conventional glass melting procedures (1200-2000 °C) [Hreniak2002, Suzuki2006, You2006, Hayakawa2005]. Additionally, there are examples of luminescent nanocrystals dispersed in sol-gel derived matrices [Counio1996, Penard2007, Bullen2004] and nanophosphors produced through the sol-gel route [Lo1999, Lin2007].

The fluorescence properties in sol-gel materials are often provided by the incorporation of transition metal and rare earth ions (Eu^{2+}, Ce^{3+}, Sm^{3+}, Eu^{3+}, Dy^{3+}, Er^{3+} and Tm^{3+}) [Cordoncillo2001, Cordoncillo2002, Escribano2008, Julián2004, Nobre2008, de Zea Bermudez1998, Li1998, Ribeiro1998, Han2003, Campostrini1992, Zhang2004, Li1998]. In particular, the Eu^{2+} has unique properties and has been used in many systems, but usually high processing temperatures are required for the stabilization of the divalent state. Cordoncillo et al. presented the first work in this field where the Eu^{2+} is incorporated in the sol-gel polysiloxane at room temperature [Cordoncillo1998]. Nevertheless, the usual route in phosphors and glasses consists in the insertion of Eu^{3+} in the matrix and its reduction to Eu^{2+} with heat-treatment. An example of luminescent silicon oxycarbide glass derived from sol-gel, doped with Eu^{3+} and reduced to Eu^{2+} by means of heat-treatment, was published by Zhang et al. in 2004 [Zhang2004]. The employment of UV irradiation can be an alternative reduction route [Suzuki2006]. Often the europium is included in a crown ether complex [Suzuki2006].

As mentioned before, sol-gel glasses can be simple dry monoliths or be heat-treated at different temperatures [Lin2000, Cordoncillo1998, Garcia1995, Green1997, Li1998, Li2000b, Martucci2003, You2006, Hayakawa2005]. The luminescent materials obtained through the sol-gel method are usually SiO_2, but also TiO_2-SiO_2, SiOC glasses, etc [You2006, Zhang2004, Das2007, Das2008, Karakuscu2009a, Karakuscu2009b].

Besides the doped sol-gel materials, luminescence properties have been observed to arise also in non-doped sol-gel glasses [Green1997, Garcia1995, Lin2000]. The photoluminescence properties mainly originate from defects like carbon impurities [Bekiari1998, Stathatos1998], oxygen vacancies [Uchino2000, Yang2001, Prokes1998, Lopez2000, Li1998], silicon dangling bonds [Baran2004, Prokes1998, Lopez2000], non-bridging oxygen defects [Yang2001], radical carbonyl defects [Hayakawa2003], $SiOH^-$ radicals [Hreniak2002] and charge transfer process between the silicon and the oxygen atom [Han2003, Han2002, Garcia1995, Green1997]. Nevertheless, the origin of the luminescence in sol-gel materials has not been yet completely clarified.

Examples of undoped sol-gel materials heat-treated in air at different temperatures (200-500 °C) were published [Green1997, Garcia1995, Lin2000]. Green *et al.* reported that the photoluminescence properties were related to the carbon impurities present in the glasses, similarly to the papers reported by Pivin *et al.* for irradiated polysiloxanes [Pivin1998, Pivin2000], and in agreement to our results. Analogously to the materials treated in the present thesis, as the annealing temperature increases up 500-700 °C, the sol-gel polymers turn an opaque brown-black color, due to the formation of free carbon [Green1997, Garcia1995]. Lin *et al.* highlighted the red-shift of the excitation and emission spectra after increasing heat-treatment [Lin2000]. The publications of Green, Garcia and Lin *et al.* represent analyses of the photoluminescence properties of sol-gel materials observed at different heat-treatment temperatures, with similar behavior as the heat-treated silicon-based polymers embraced in this thesis. Nevertheless, a detailed analysis at increasing temperature steps is not included [Green1997, Garcia1995, Lin2000].

A specific class of luminescent sol-gel materials is constituted by the inorganic–organic hybrid materials, where the siliceous domains or siloxane networks are alternated to organic moieties, which show luminescent properties both in the doped and undoped state [Stathatos2003, Carlos2001, Sá Ferreira1999, Sá Ferreira2006, Carlos1999, de Zea Bermudez1999, Fu2004, Carlos2004, Han2002, Sanchez1994, Bekiari2001, Bekiari2000, Brankova2003, Stathatos2000, Cordoncillo2001, Cordoncillo2002, Escribano2008, Julián2004, Nobre2008, de Zea Bermudez1998, Ribeiro1998, Han2003]. Depending on the chemical structure, the luminescence properties were attributed to the presence of oxygen defects, to the photoinduced proton-transfer between defects as NH^+ and NH^- [Carlos2004], matrix defects, silicon dangling bonds, peroxy radicals, adsorbed oxygen radical, non-bridging oxygen hole centers [Cordoncillo2001].

Thus, in general, the intrinsic luminescence properties of undoped sol-gel materials were attributed to the presence of various defects.

Photoluminescence emission was detected also in silsesquioxanes $R_8(SiO_{1.5})_8$ and was attributed to the charge-transfer transition from the silsesquioxane cage to the ligands, which are responsible for a monomer fluorescence and an exciplex transition [Azinovic2002]. Furthermore, cubic silsesquioxanes were reported to tune the luminescence properties of their substituents, which may suggest 3D conjugation in the excited state [Laine2008]. Undoped urea-bridged silsesquioxanes were also reported to be intensely luminescent [Gomez2008]. In addition, chemical incorporation of silsesquioxanes in OLED devices can improve efficiency and lifetime [Chan2009].

4.3. References

[Azinovic2002] D. Azinovic, J. Cai, C. Eggs, H. König, H.C. Marsmann, S. Veprek, *J. Lumin.*, 97 *40–50*, 2002

[Baran2004] M. Baran, B. Bulakh, N. Korsunska, L. Khomenkova, J.Jedrezejewski, *Eur. Phys. Appl. Phys.*, 27, *285-287*, 2004

[Bekiari1998] V. Bekiari, P. Lianos, *Langmuir*, 14 [13], 1998

[Bekiari2000] V. Bekiari, P. Lianos, U. Lavrencic Stangar, B. Orel, P. Judeinstein, *Chem. Mater.*, 12, *3095-3099*, 2000

[Bekiari2001] V. Bekiari, E. Stathatos, P. Lianos, U. L. Stangar, B. Orel, P. Judeinstein, *Monatsh. Chem.*, 132, *97-102*, 2001

[Belfield1998] K. D. Belfield, C. Chinna, O. Najjar, *Macromolecules*, 31, *2918-2924*, 1998

[Bernard2006] S. Bernard, K. Fiaty, D. Cornu, P. Miele, P. Laurent, *J. Phys. Chem. B,* 110, *9048-9060*, 2006

[Birkofer1962] L. Birkofer, A. Ritter, F. Richter, *Tetrahedron Lett.*. 195, 1962

[Birot1995] M. Birot, J.-P. Pillot, J. Dunoguès, *Chem. Rev.*, 95, *1443-1477*, 1995

[Bisberg1995] J. Bisberg, W. J. Cumming, R. A. Gaudiana, K. D. Hutchinson, R. T. Ingwall, E. S. Kolb, P. G. Metha, R. A. Minns, C. P. Petersen, *Macromolecules,* 28, *386-389*, 1995

[Brankova2003] T. Brankova, V. Bekiari, P. Lianos, *Chem. Mater.* 15, *1855-1859*, 2003

[Brequel2004] H. Bréquel, J. Parmentier, S. Walter, R. Badheka, G. Trimmel, S. Masse, J. Latournerie, P. Dempsey, C. Turquat, A. Desmartin-Chomel, L. Le Neindre-Prum, U. A. Jayasooriya, D. Hourlier, H.-J. Kleebe, G. D. Sorarù, S. Enzo, F. Babonneau, *Chem. Mater.* 16, *2585-2598*, 2004

[Bullen2004] C. Bullen, P. Mulvaney, C. Sada, M. Ferrari, A. Chiasera, A. Martucci, *J. Mater. Chem.*, 14, *1112–1116*, 2004

[Campostrini1992] R. Campostrini, G. Carturan, M. Ferrari, M. Montagna, O. Pilla, *J. Mater. Res.*, 7[3], 1992

[Carlos1999] L. D. Carlos, V. de Zea Bermudez, R. A. Sá Ferreira, L. Marques, M. Assunção, *Chem. Mater.*, 11, *581-588*, 1999

[Carlos2001] L. D. Carlos, R. A. Sá Ferreira, V. de Zea Bermudez, S. J. L. Ribeiro, *Adv. Funct. Mater.* 11 [2], *111-115*, 2001

[Carlos2004] L. D. Carlos, R. A. Sá Ferreira, R. N. Pereira, M. Assunção, V. de Zea Bermudez, *J. Phys. Chem. B,* 108, *14924-14932*, 2004

5. Experimental procedure

[Chan2009] K. L. Chan, P. Sonar, A. Sellinger, *J. Mater. Chem.*, 2009, Feature Article

[Chojnowski2000] J:Chojnowski, M. Cypryk, *Polysiloxanes*, Ch 1 in: R. G. Jones, W. Ando, J. Chojnowski, *Silicon-Containing Polymers - The Science and Technology of Their Synthesis and Applications*, Kluwer Academic Publishers, 2000

[Colombo1999a] P. Colombo, M. Modesti *J. Am. Ceram. Soc.*, 82 [3], *573–578*, 1999

[Colombo1999b] P. Colombo, M. Modesti, *J. Sol-Gel Sci. Techn.*, 14, *103–111*, 1999

[Colombo2001] P. Colombo, J. R. Hellmann, D. L. Shelleman *J. Am. Ceram. Soc.*, 84 [10], *2245–2251*, 2001

[Colombo2004] P. Colombo, E. Bernardo, L. Biasetto *J. Am. Ceram. Soc.* 87 [1], *152-154*, 2004

[Cordoncillo1998] E. Cordoncillo,a B. Viana,b P. Escribanoa, C. Sanchez, *J. Mater. Chem.*, 8 [3], *507–509*, 1998

[Cordoncillo2001] E. Cordoncillo, J. Guaita, P. Escribano, C. Philippe, B. Viana, C. Sanchez, *Opt. Mater.*, 18 *309-320*, 2001

[Cordoncillo2002]E. Cordoncillo, P. Escribano, J. Guaita, C. Philippe, B. Viana, Clement Sanchez, *J. Sol-Gel Sci. Techn.*, 24, *155–165*, 2002

[Counio1996] G. Counio, S. Esnouf, T. Gacoin, J.-P. Boilot, *J. Phys. Chem.* 100, *20021-20026*, 1996

[Das2007] G. Das, P. Bettotti, L. Ferraioli, R. Raj, G. Mariotto, L. Pavesi, G. D. Soraru, *Vib. Spectrosc.*, 45 [1], 61-68, 2007

[Das2008] G. Das, L. Ferraioli, P. Bettotti, F. De Angelis, G. Mariotto, L. Pavesi, E. Di Fabrizio, G. D. Soraru, *Thin Solid Films*, 516 [20], 6804-6807, 2008

[de Zea Bermudez1998] V. de Zea Bermudez, L. D. Carlos, M. C. Duarte, M. M. Silva, C. J. R. Silva, M. J. Smith, M. Assunção, L. Alcácer, *J. Alloy. Compd.*, 275–277, *21–26*, 1998

[de Zea Bermudez1999] V. de Zea Bermudez, L. D. Carlos, L. Alcácer, *Chem. Mater.*, 11, *569-580*, 1999

[Dias2000] F. B. Dias, J. C. Lima, A. l. Maçanita, S. J. Clarson, A. Horta, I. F. Piérola, *Macromolecules*, 33, *4772-4779*, 2000

[Dias2002] F. B. Dias, J. C. Lima, A. Horta, I. F. Piérola, A. L. Maçanita, *Macromolecules*, 35, *7082-7088*, 2002

[Du2006] X.-W. Du, Y. Fu, J. Sun, P. Yao, *J. Appl. Phys.*, 99, *935031-4*, 2006

[Ebsworth1961] A. V. Ebsworth, M. J. Mays, *J. Chem. Soc.*, *4879-4882*, 1961

[Ebsworth1962] A. V. Ebsworth, M. J. Mays, *Angew. Chem.*, 74, *113*, 1962

[Escribano2008] P. Escribano, B. Julián-López, J. Planelles-Aragó, E. Cordoncillo, B. Viana, C. Sanchez, *J. Mater. Chem.*, 18, *23–40*, 2008

[Ferraioli2008] L. Ferraioli, D. Ahn, A. Saha, L. Pavesi, R. Raj, *J. Am. Ceram. Soc.*, 91 [7], *2422–2424*, 2008

[Fogarty2003] H. A. Fogarty, D. L. Casher, R. Imhof, T. Schepers, D. W. Rooklin, J. Michl, *Pure Appl. Chem.*, 75 [8], *999-1020*, 2003

5. Experimental procedure

[Fu2004] L. Fu, R. A. Sa´ Ferreira, N. J. O. Silva, L. D. Carlos, V. de Zea Bermudez, J. Rocha, *Chem. Mater.*, 16, *1507-1516*, 2004

[Fukushima2004] M. Fukushima, E. Yasuda, Y. Teranishi, K. Nakamura, Y. Tanabe, *J. Ceram. Soc. Jpn.*, 112 [11], *612-614*, 2004

[Garcia1995] J. Garcia M., M.A. Mondragon, C. Tellez S., A. Campero, V.M. Castano, *Mater. Chem. Phys.*, 41, *15-17*, 1995

[Gomez2008] M. L. Gomez, D. P. Fasce, R. J. J. Williams, C. M. Previtali, L. Matejka, J. Plestil, J. Brus, *Macromol. Chem. Phys.*, 209, *634–642*, 2008

[Gorbatenko1977] V. I. Gorbatenko, M. N. Gertsyuk, L. I. Samarai, *Zh. Org. Khim.* 13, *899*, 1977

[Green1997] W. H. Green, K. P. Le, J. Grey, T. T. Au, M. J. Sailor, *Science,* 276, *1826*, 1997

[Gumula2004] T. Gumula, C. Paluszkiewicz, M. Blazewicz, *J. Mol. Struct.*, 704, *259–262*, 2004

[Gvishi1997] R. Gvishi, U. Narang, G. Ruland, D. N. Kumar, P. N. Prasad, *Appl. Organomet. Chem.*, 11, *107–127*, 1997

[Hamanishi1993] K. Hamanishi, H. Shizuka, *J. Chem. Soc. Faraday Trans.*, 89 [16], *3007*, 1993

[Han2002] Y. Han, J. Lin, H. Zhang, *Mater. Lett.*, 54, *389– 396*, 2002

[Han2003] Y. Han, J. Lin, *J. Solid State Chem.*, 171, *396-400*, 2003

[Harshe2004a] R. R. Harshe, C. Balan, R. Riedel, *J. Eur. Ceram. Soc.*, 24, *3471–3482*, 2004

[Harshe2004b] R. R. Harshe, PhD thesis, TU Darmstadt, 2004

[Hayakawa2003] T.Hayakawa, A. Hiramitsu, M. Nogami, *Appl. Phys. Lett.*, 82, *18*, 2003

[Hayakawa2005] T. Hayakawa, M. Nogami, *Sci. Technol. Adv. Mat.*, 6, *66–70*, 2005

[Hennecke1988] M. Hennecke, P. Strohriegl, *Makromol. Chem.*, 189, *2601-2609*, 1988

[Hreniak2002] D. Hreniak, M. Jasiorski, K. Maruszewski, L. Kepinski, L. Krajczyk, J. Misiewicz, W. Strek *J. Non-Cryst. Solids,* 298, *146–152*, 2002

[Hua1997] C. Hua Yuan, S. Hoshino, S. Toyoda, H. Suzuki, M. Fujiki, N. Matsumoto, *Appl. Phys. Lett.,* 71 [23], 1997

[Ionescu2009a] E. Ionescu, C. Gervais, F. Babonneau, *The Polymer-to-Ceramic Transformation*, Ch 2. in *Polymer-derived-ceramics: Theory and Applications*, Ed. by P. Colombo, R. Riedel, G.D. Sorarù, H.-J. Kleebe, DESTech Publications, Inc. (September 28, 2009), Lancaster, PA, USA

[Itoh1996] T. Itoh, M.-H. Yang, C. Chou, *J. Chem. Soc., Faraday Trans.,* 92 [19], *3593-3597*, 1996

[Itoh2001] T. Itoh *Res. Chem. Intermed.*, 27 [6], *669–685*, 2001

[Itoh2002] T. Itoh, M.-H. Yang, *J. Polym. Sci. Pol. Phys.*, 40, *854–861*, 2002

[Julián2004] B. Julián, R. Corberán, E. Cordoncillo, P. Escribano, B. Viana, C. Sanchez, *J. Mater. Chem.*, 14, *3337–3343*, 2004

5. Experimental procedure

[Karakuscu2009a] A. Karakuscu, R. Guider, L. Pavesi, G. D. Sorarù, *Nanostructured Materials and Nanotechnology II*, Edited by Sanjay Mathur and Mrityunjay Singh, The American Ceramic Society, *85-91*, 2009

[Karakuscu2009b] A. Karakuscu, R. Guider, L. Pavesi, G. D. Sorarù, *J. Am. Ceram. Soc.*, 92 [12], *2969-2974*, 2009

[Kim1997] H. K. Kim, M.-K. Ryu, S.-M. Lee, *Macromolecules*, 30, *1236-1239*, 1997

[Kim2005] C. Kim, K. Kwark, C.-G. Song, *Appl. Organometal. Chem.*, 19, *108-112*, 2005

[Klebe1966] J. F. Klebe, J. G. Murray, *US Patent 3 352 799*, 1966

[Kleebe2006] H.- J. Kleebe, G. Gregori, F. Babonneau, Y. D. Blum, D. B. Macqueen, S. Masse, *Int. J. Mater. Res.*, 97 [6], *699-709*, 2006

[Kroke2000] E. Kroke, Y.-L. Li, C. Konetschny, E. Lecomte, C. Fasel, R. Riedel, *Mater. Sci. Eng.*, 26, 97-199, 2000

[Laine2008] S. Sulaiman, A. Bhaskar, J. Zhang, R. Guda, T. Goodson III, R. M. Laine, *Chem. Mater.*, 20, *5563–5573*, 2008

[Landenburg1872] Ladenburg, *Liebigs Ann. Chem.*, 164, *300*, 1872

[Latournerie2006] J. Latournerie, P. Dempsey, D. Hurlier-bahloul, J.-P. Bonnet, *J. Am. Ceram. Soc.*, 89 [5], *1485-1491*, 2006

[Lee1999] L.-L. Lee, D.-S. Tsai, *J. Am. Ceram. Soc.*, 82 [10], *2796-800*, 1999

[Li1998] Y. H. Li, C. M. Mo, L. Z. Yao, R. C. Liu, W. L. Cai, X. M.. Li, Z. Q. Wu, L. D. Zhang, *J. Phys.: Condens. Matter*, 10, *1655–1664*, 1998

[Li2000b] G. Li, L. W. Burggraf, J. R. Shoemaker, D. Eastwood, A. E. Stiegman, *Appl. Phys. Lett.*, 76[23], *3373-3375*, 2000

[Li2001] Y. L. Li, E. Kroke, R. Riedel, C. Fasel, C. Gervais, F. Babonneau, *Appl. Organometal. Chem.*, 15, *820–832*, 2001

[Liemersdorf2008] S. Liemersdorf, R. Riedel, J. Oberle, *J. Am. Ceram. Soc.*, 91 [1], *325–328*, 2008

[Lin2000] J. Lin, K. Baerner, *Mater. Lett.*, 46, *86–92*, 2000

[Lin2007] C.-C. Lin, K.-M. Lin, Y.-Y. Li, *J. Lumin.*, 126 [2], *795-799*, 2007

[Lo1999] J.-R. Lo, S. Ezhilvalavan, T.-Y. Tseng, *Jpn. J. Appl. Phys.*, 38, *1390-1393*, 1999

[Lopez2000] N. Lopez, F. Illas, G. Pacchioni, *J. Phys. Chem. B*, 104, *5471-5477*, 2000

[Ma2007] N. Ma, Y. Liao, Z. Sun, S. Huang, *Mater. Lett.*, 61, *4163-4165*, 2007

[Maçanita1991] A. L. Maçanita, I. F. Piérola, A. Horta, *Macromolecules*, 24, *1293-1298*, 1991

[Maçanita1994a] A. L. Maçanita, A. Horta, I F. Piérola, *Macromolecules*, 27, *958-963*, 1994

[Maçanita1994b] A. L. Maçanita, A. Horta, I. F. Piérola, *Macromolecules*, 27, *3797-3803*, 1994

5. Experimental procedure

[Maçanita1994c] A. L. Maçanita, P. Danesh, F. Peral, A. Horta, I. F. Piérola, *J. Phys. Chem.* 98, *6548-6551*, 1994

[Martucci2003] A. Martucci, A. Chiasera, M. Montagna, M. Ferrari, *J. Non-Cryst. Solids*, 322, *295–299*, 2003

[Mera2009a] G. Mera, R. Riedel, *Organosilicon-Based Polymers as Precursors for Ceramics*, Ch 2.1.3. in *Polymer-derived-ceramics: Theory and Applications*, Ed. by P. Colombo, R. Riedel, G.D. Sorarù, H.-J. Kleebe, DESTech Publications, Inc. (September 28, 2009), Lancaster, PA, USA

[Mera2009b] G. Mera, R. Riedel, F. Poli, and K. Müller, *J. Eur. Ceram. Soc.*, 29, *2873-2883*, 2009

[Michl1988] J. Michl, J. W. Downing, T. Karatsu, A. J. McKinley, G. Poggi, G. M. Wallraff, R. Sooriyakumaran, R. D. Miller, *Pure Appl. Chem.*, 60 [7], *959-972*, 1988

[Morcos2008a] R. M. Morcos, G. Mera, A. Navrotsky, T. Varga, R. Riedel, F. Poli, K. Müller, *J. Am. Ceram. Soc.*, 91 [10], *3349–3354*, 2008

[Nagai1996] K. Nagai, K. Utsunomiya, N. Takamiya, N. Nemoto, M. Kaneko, *J. Polym. Sci. Polym. Phys.* 34, *2059*, 1996

[Nespurek2002] S. Nespurek, A. Kadashchuk, Y. Skryshevski, A. Fujii, K. Yoshino, *J. Lumin.*, 99 *131-140*, 2002

[Nobre2008] S. S. Nobre, C. D. S. Brites, R. A. S. Ferreira, V. de Zea Bermudez, C. Carcel, J. J. E. Moreau, J. Rocha, M. Wong Chi Man, L. D. Carlos, *J. Mater. Chem.*, 18, *4172–4182*, 2008

[Penard2007] A.-L. Penard, T. Gacoin, J.-P. Boilot, *Acc. Chem. Res.*, 40, *895-902*, 2007

[Perry1990] R. J. Perry, T. Ching, *Patent US4950950 (A)*, 1990

[Pivin1998] J. C. Pivin, P. Colombo, M. Sendova-Vassileva, J. Salomon, G. Sagon, A. Quaranta, *Nucl. Instrum. Meth. B*, 141, *652-662*, 1998

[Pivin2000] J.C. Pivin, M. Sendova-Vassileva, P. Colombo, A. Martucci, *Mat. Sci. Eng. B*, 69–70, *574–577*, 2000

[Prokes1998] S. M. Prokes, W.E.Carlos, S.Veprek, Ch.Ossadnik, *Phys. Rev. B*, 58, *23*, 1998

[Pump1962a] J. Pump, U. Wannagat, *Angew. Chem.* 74, *117*, 1962

[Pump1962b] J. Pump, U. Wannagat, *Justus Liebigs Ann. Chem.*, 652, *21-27*, 1962

[Pump1963] J. Pump, E. G. Rochow and U. Wannagat, *Monatsh. Chem.*, 94, *588-598*, 1963

[Radovanovic1999] E. Radovanovic, M.F. Gozzi, M.C. Goncalves, I.V.P. Yoshida, *J. Non-Cryst. Solids*, 248, *37-48*, 1999

[Rathore2009] J. S. Rathore, L. V. Interrante, *Macromolecules*, 42, *4614–4621*, 2009

[Razuvaev1987] G. A. Razuvaev, A. S. Gordetsov, A. P. Kozina, T. N. Brevnova, V. V. Semenov, S. E. Skobeleva, N. A. Boxer, Yu. I. Dergunov, *J. Organomet. Chem.* 327, *303*, 1987

[Ribeiro1998] S.J.L. Ribeiro, K. Dahmouche, C.A. Ribeiro, C.V. Santilli, S.H. Pulcinelli, *J. Sol-Gel Sci. Techn.*, 13, *427–432*, 1998

5. Experimental procedure

[Riedel1998] R. Riedel, E. Kroke, A. Greiner, A. O. Gabriel, L. Ruwisch, J. Nicolich, *Chem. Mater.*, 10, *2964-2979*, 1998

[Riedel2006] R. Riedel, G. Mera, R. Hauser, A. Klonczynski, *J Ceram Soc Jpn*, 114, *425-444*, 2006

[Rooklin2003] D. W. Rooklin, T. Schepers, M. K. Raymond-Johansson, J. Michl, *Photochem. Photobiol. Sci.*, 2, *511-517*, 2003

[Sá Ferreira1999] R. A. Sá Ferreira, L.D. Carlos, V. de Zea Bermudez, *Thin Solid Films*, 343-344, *476-180*, 1999

[Sá Ferreira2006] R.A. Sá Ferreira, A.L. Ferreira, L.D. Carlos, *J. Non-Cryst. Solids*, 352, *1225–1229*, 2006

[Saha2006] A. Saha, R. Raj, D. L. Williamson, *J. Am. Ceram. Soc.*, 89 [7], *2188–2195*, 2006

[Saha2007] A. Saha, R. Raj, *J. Am. Ceram. Soc.*, 90 [2], *578–583*, 2007

[Sacarescu2008] L. Sacarescu, I. Mangalagiu, M. Simionescu, G. Sacarescu, R. Ardeleanu, *Macromol. Symp.*, 267, *123-128*, 2008

[Salom1987a] C. Salom, A. Horta, I. Hernández-Fuentes, I. F. Piérola, *Macromolecules*, 20, *696-698*, 1987

[Salom1987b] C. Salom, M. R. Gomez-Anton, A. Horta, I. Hernández-Fuentes, I. F. Piérola, *Macromolecules*, 20, *1627-1630*, 1987

[Salom1989] C. Salom, I. Hernindez-Fuentes, I. F. Piérola, A. Horta, *Macromolecules*, 22, *1874-1878*, 1989

[Sanchez1994] C. Sanchez, B. Lebeau, F. Chaput, J.-P. Boilot, *Adv. Mater.*, 15 [23], *1969-1994*, 2003

[Sanchez2008] J. C. Sanchez, W. C. Trogler, *Macromol. Chem. Phys.* 209, *1527–1540*, 2008

[Schubert2004] U. Schubert, N. Hüsing, *Synthesis of Inorganic Materials*, Wiley-VCH, Weinheim, Ch. 5, 2004

[Sharma2006] A. Sharma, M. Katiyar, Deepak, S. Seki, S. Tagawa, *Appl. Phys. Lett.* 88, *143511*, 2006

[Sharma2007] A. Sharma, M. Katiyar, Deepak, S. Seki, *J. Appl. Phys.* 102, *084506*, 2007

[Shaw1983] J. M. Shaw, M. Hatzakis, J. Paraszczak, J. Liutkus, E. Babich, *Polym. Eng. Sci.*, 23 [18], 1983

[Skryshevskii2003] Y. A. Skryshevskii, *J. Appl. Spectrosc.*, 70, *6*, 2003

[Singh2008] R. Singh, M. Katiyar, Mater. *Res. Soc. Symp. Proc.*, 1091, 2008

[Soraru1994] G. D. Sorarù, *J. Sol-Gel Sci. Techn.*, 2, *843-848*, 1994

[Soum2000] A. Soum, *Polysilazanes*, Ch 11 in: R. G. Jones, W. Ando, J. Chojnowski, (Eds.), *Silicon-Containing Polymers - The Science and Technology of Their Synthesis and Applications,* Kluwer Academic Publishers, 2000

[Stathatos2000] E. Stathatos, P. Lianos, U. Lavrencic-Stangar, B. Orel, P. Judeinstein, *Langmuir,* 16, *8672-8676*, 2000

[Stathatos2002] E. Stathatos, P. Lianos, U. Lavrencic-Stangar, B. Orel, *Adv. Mater.*, 14 [5], *354-357*, 2002

[Stathatos2003] E. Stathatos, P. Lianos, B. Orel, A. Surca Vuk, R. Jese, *Langmuir*, 19, *7587-7591*, 2003

[Sun1991] Y.-P. Sun, R. D. Miller, R. Sooriyakumaran, J. Michl, *J. Inorg. Organomet. P.*, 1 [1], 1991

[Suzuki1996] H. Suzuki, *Adv. Mater.*, 8, *657-659*, 1996

[Suzuki1997] M. Suzuki, Y. Nakata, H. Nagai, T. Okutani, N. Kushibiki, M. Murakami, *Mater. Sci. Eng. B*, 49, *172-174*, 1997

[Suzuki1998] H. Suzuki, S. Hoshino, C.-H. Yuan, M. Fujiki, S. Toyoda, N. Matsumoto, *Thin Solid Films*, 331, *64-70*, 1998

[Suzuki2000] H. Suzuki, S. Hoshino, K. Furukawa, K. Ebata, C. Hua Yuan, I. Bleyl, *Polym. Adv. Technol.*, 11, *460-467*, 2000

[Suzuki2006] R. Suzuki, S. Takei, E. Tashiro, K.-I. Machida, *J. Alloy. Compd.*, 408–412, *800–804*, 2006

[Szadkowska-Nicze2004] M. Szadkowska-Nicze, J. Mayer, *J. Polym. Sci. A1*, 42, *6125–6133*, 2004

[Todesco1986] R. V. Todesco, P. V. Kamat, *Macromolecules*, 19, *196-200*, 1986

[Toulokhonova2003] I. Toulokhonova, B. Bjerke-Kroll, R. West, *J. Organomet. Chem.*, 686, *101-104*, 2003

[Trimmel2003] G. Trimmel, R. Badheka, F. Babonneau, J. Latournerie, P. Dempsey, D. Bahloul-Houlier, J. Parmentier, G. D. Sorarù, *J. Sol-Gel Sci. Techn.*, 26, *279–283*, 2003

[Uchino2000] T. Uchino, M. Takahashi, T. Yoko, *Phys. Rev. B*, 62 [5], *2983-2986*, 2000

[Vakifahmetoglu2009] C. Vakifahmetoglu, I. Menapace, A. Hirsch, L. Biasetto, R. Hauser, R. Riedel, P. Colombo, *Ceram. Int.*, 35, *3281-3290*, 2009

[Vinod2003] M. P. Vinod, D. Bahnemann, P.R. Rajamohanan, K. Vijayamohanan, *J. Phys. Chem. B*, 107, *11583-11588*, 2003

[Wallraff1992] G. M. Wallraff, M. Baier, A. Diaz, R. D. Miller, *J. Inorg. Organomet. P.*, 2 [1], 1992

[Wang2005] H. Wang, S.-Y. Zheng, X.-D. Li, D.-P. Kim, *Micropor. Mesopor. Mat.*, 80, *357–362*, 2005

[Watanabe2001] A. Watanabe, T. Sato, M. Matsuda, *Jpn. J. Appl. Phys.*, 40, *6457-6463*, 2001

[Watanabe2003] A. Watanabe, *J. Organomet. Chem.*, 685, *122-133*, 2003

[Yang2001] P. Yang, C. F. Song, M. K. Lü, J. Chang, Y. Z. Wang, Z. X. Yang, G. J. Zhou, Z. P. Ai, D. Xu, D. L. Yuan, *J. Solid State Chem.*, 160, *272-277*, 2001

[You2006] H. You, M. Nogami, *J. Alloy. Compd.*, 408-412, *796-799*, 2006

[Zhang2004] Y. Zhang, A. Quaranta, G. D. Sorarù, *Opt. Mater.*, 24, *601–605*, 2004

[Zhou2008] Q. Zhou, S. Yan, C. C. Han, P. Xie, R. Zhang, *Adv. Mater.* 20, *2970–2976*, 2008

5. Experimental procedure

In this chapter, the characterization methods used in the present thesis work are described. For the less common characterization methods, an explanation of the principles of the method is also supplied. Moreover, a short resume of the materials employed is presented. For the commercially available materials only the applied heat-treatment is described.

5. 1. Starting materials

Commercially available polymethylsilsesquioxane Wacker-Belsil® PMS MK (Wacker Chemie AG, Munich, Germany), $(CH_3\text{-}SiO_{3/2})_x$, polydimethylsiloxane PDMS (SIGMA), polyvinylmethylsiloxane PVMS homopolymer (ABCR), vinylmethylsiloxane VMS linear homopolymer (ABCR), polyureamethylvinylsilazane (PUMVS, Ceraset^{TM}) (KiON Corp., Clariant, USA) $(\text{-}NH(CH_3)SiH\text{-})_{0.77}(\text{-}(NRC\text{=}O)_{0.01}(NH(HC\text{=}CH_2)SiCH_3\text{-})_{0.22}$, polysilazane VL20 or KiON Ceraset Polysilazane 20 (KiON Corp.) and KiON S (KiON Corp.) were used as precursors.

Four polysilylcarbodiimides (S1, S2, S3 and S4) were synthesized by the reaction of diphenyldichlorosilane (>98%, Fluka), phenylmethyldichlorosilane (97%, ABCR), phenyldichlorosilane (97%, ABCR) and phenylvinyldichlorosilane (95%, ABCR), respectively, with bistrimethylsilylcarbodiimide, using pyridine (anhydrous, 99.8%, Sigma Aldrich) as catalyst. First, bistrimethylsilylcarbodiimide (0.047 mol) was mixed under stirring with pyridine (0.024 mol) and then the substituted dichlorosilane (0.047 mol) was added. The reaction mixture was kept under reflux at 66 °C for 6 h (44 °C for S3 since the boiling point of PhHSiCl$_2$ is 65–66 °C) and subsequently at 120 °C for another 12 h (66 °C for S3). The formation of the substituted polysilylcarbodiimide was monitored by means of ^{29}Si NMR spectroscopy. After completion of the reaction, the by-product trimethylchlorosilane was removed by distillation. The yield of the reaction is about 80%. Bistrimethylsilylcarbodiimide was synthesized using dicyandiamide (C$_2$H$_4$N$_4$, SKW Trostberg), 1, 1, 1,

3, 3, 3-hexamethyldisilazane (Merck) and ammonium sulfate $(NH_4)SO_4$ (>99%, Sigma Aldrich) as catalyst, according to the well known reaction: $H_2N-C(=NH)-(NH)-CN + 2Me_3SiN(H)SiMe_3 \rightarrow ((NH_4)_2SO_4)\uparrow \rightarrow 2Me_3SiNCNSiMe_3 + 2NH_3\uparrow$.

The copolymer GV was obtained by the reaction of commercial VL20 (10 g) with the synthesized polydiphenylsilylcarbodiimide S1 (10 g) (1:1 wt%), by simple stirring under argon for one day at room temperature.

All reactions were carried out under inert argon atmosphere using standard Schlenk techniques. Air and/or hydrolysis-sensitive polymers were handled in glove box under argon. The joint grease was silicon-free (KWS).

5.2. Heat-treatment

In all heat-treatments, the samples were prepared by thermolysis of 1-2 g of the preceramic polymer in a quartz crucible placed in a quartz tube (h = 50 cm, d_i= 3 cm) under a steady flow of purified argon (50 mL/min) in a programmable horizontal Al_2O_3 tube furnace (Gero - Öfen, Typ 40-300/7 85). The samples were annealed with a heating rate of 50 °C/h from room temperature to the target temperature (200, 300, 400, 500, 600 and 700 °C). The dwelling time was 2 h at the final annealing temperature, followed by free cooling down to room temperature. In order to avoid the oxygen contamination, the quartz tube was sealed with silicon-free joint grease (KWS) and was evacuated and refilled with argon three times. The decomposition gases together with the argon exit the system passing through two exhaust bottles partially filled with glycerine. This prevents entrance of air in the reverse direction inside the quartz tube.

5.3. Photoluminescence measurements

5.3.1. Photoluminescence spectrometry

The photoluminescence measurements were recorded using a commercial Cary Eclipse Varian spectrophotometer using a Xenon lamp as the excitation source, sensitive across the wavelength range 200-800 nm. Both fluorescence and phosphorescence emission spectra and excitation spectra were recorded.

Fluorescence is an electric-dipole allowed transition and shows lifetimes of $\tau \approx 10^{-7}\text{-}10^{-8}$ s, while phosphorescence is an electric-dipole forbidden transition with lifetimes of about $\tau \approx$ a few 10^{-3} s. Thus, electric-dipole forbidden transitions have longer lifetimes than electric-dipole allowed transitions. In phosphorescence measurements the detection time is delayed in respect to fluorescence measurements, in order to detect forbidden transitions, which are not detectable in fluorescence measurements. All samples were subjected to phosphorescence measurements, with excitation wavelength 250 nm, using the following parameters:

- delay time: 0.2 ms
- gate time: 5 ms
- total decay time: 0.020 s (20 ms)
- PMT voltage: 1000 V.

The fluorescence emission and excitation spectra were recorded for several excitation and emission wavelengths, in order to get the maximum excitation or emission peaks. Furthermore, the fluorescence emission spectra were always recorded using 250 and 360 nm as excitation wavelengths. All measurements were performed in air at room temperature and using comparable instrument parameters. For the polysilylcarbodiimides samples prepared in glove box and sealed, the measurements were performed in argon. The samples were measured in powder or liquid state in 1 mm precision cells made of Quartz Suprasil®, 100-QS, Hellma®. The cell was positioned at 35° of angle in order to minimize the scattering and maximize the emission intensity.

The original spectra were corrected by subtracting the effect of light scattering. The spectrum of the scattering was obtained by stimulating a highly refractive and non-luminescent material with the same parameters used for the samples. Moreover, the corrected spectrum was calibrated with an experimental calibration spectrum, as the detection power of the spectrometer at longer wavelengths is weak.

When a second band reproducing the shape of the 1st peak is observable in the spectra at longer wavelengths, the measurements were repeated with the same parameters and by covering the main emission using suitable filters (OG 570 and RG 610). In absence of the main peak, no emission was detected in the orange-red range. Therefore, this band was confirmed to be the second order peak of the main emission and not a true emission, and was eliminated from the spectrum.

The fluorescence spectra of samples treated at different temperatures or after different times of exposure to air were compared on the same graphs. The intensities of the spectra are comparable within an error of about 20 %, due to the different powder packing and the exact position of the quartz cell.

The fluorescence spectra were deconvoluted in energy scale using a Gaussian fit in MathWorks (Matlab), Inc. software. The fluorescence intensity was calculated as the integral of the emission spectrum in energy scale.

In order to convert the spectral emissions from the photoluminescence measurements into colors perceived by the human eye, the CIE (International Commission on Illumination) chromaticity diagram 1931 was used as color space [CIE]. In fact, to plot all visible colors a 3D figure would be necessary. However, the concept of color can be divided into brightness and chromaticity. The chromaticity coordinates were calculated from the fluorescence spectra using a suitable program (private).

5.3.2. Lifetime measurements

Lifetime measurements were carried out on the Cary Eclipse Varian (resolution of 0.2 ms) and on a self-made machine (resolution of 100 ns).

5.3.3. Quantum efficiency

Quantum efficiency measurements were carried out with an integrating sphere on the Fluorolog III UV-Vis-NIR-Fluorescence spectrometer (HORIBA Jobin Yvon, München). As the quantum efficiency measurements are time-consuming, only a selected set of samples among all the heat-treated silicon-based polymers analyzed in this thesis was measured. The set of samples was chosen in order to cover the whole visible range. Samples with different emission ranges were measured and used as standard for the remaining samples which present the same emission range. Thus, the quantum efficiencies of the remaining samples were estimated by comparing the integral of their emission spectra (in energy scale) with the integral of the emission spectra of the samples of known quantum efficiency, with emission in the same range. Samples that emit in the UV range could not be measured. An error of 5% must be taken into account for the original measurements, which propagates in the calculation of the remaining samples.

5.4. Absorption measurements

5.4.1. UV-Visible spectroscopy

A Perkin Elmer Lambda 900 UV/VIS/NIR spectrometer was used for the optical measurements. Samples were dissolved in THF and placed in precision cells made of Quartz Suprasil®, 117.104F-QS, Hellma® (10 mm light path). The cut-off wavelength of THF is around 250 nm, thus no analyses were possible for more energetic wavelengths. All measurements were performed at room temperature.

5.4.2. Remission/Reflection measurements

Since the silicon-based polymers lose their solubility in common solvents as the annealing temperature increases, remission/reflection measurements were carried out besides absorption measurements, in order to investigate the absorption range of the heat-treated samples and its trend with increasing temperature treatments. With this method, the light remitted/reflected from the sample is taken into account rather than the light transmitted.

Considering the equation A+T+R=1, where A is the absorption, T the transmission and R the reflection, if the reflected light of a sample is detected, it is possible to calculate the sum of the absorbed and transmitted light. When the diffuse light reflection from a material is higher than the absorption, the Kubelka-Munk theory can be applied, as simplification of the more general equation A+T+R=1. The Kubelka-Munk theory is generally used for the analysis of diffuse reflectance spectra obtained from weakly absorbing samples.

The absorption/scattering ratio is calculated from the remission/reflection data with the Kubelka-Munk or remission function F(R) as follows:

$$\text{Absorption/Scattering} \approx F(R) = \frac{(1-R)^2}{2R}$$

where R is the remitted beam [Wendlandt1966, Kortüm1969].

The scattering is supposed to remain constant; therefore the absorption coefficient results from the Kubelka-Munk function. The theory is suitable for optically thick materials where more than 50 % of light is reflected and less than 20 % is transmitted. In the case of heat-treated silicon-based polymers, the majority of the samples are transparent, but after being powdered a great part of the transmission can be reduced and most of the light is reflected. The Kubelka-Munk theory is very useful to define the trend of the absorption edge of the silicon-based polymers with increasing treatment temperatures, which are not soluble anymore.

Remission is the light reflected or scattered by a material, opposed to the light transmitted or absorbed. Remission spectra were obtained by detecting all the wavelengths emitted by the material for a selected excitation wavelength, with a Perkin Elmer Lambda 900 UV/VIS/NIR spectrometer. An integrating sphere is useful to detect the radiations in all directions. Thus, it is possible to estimate, for every excitation wavelength, the amount of radiation re-emitted, and for difference, the amount absorbed.

Strictly speaking, the remission measurements do not give an accurate absorption spectrum for luminescent samples. Since the sphere does not detect selectively only the radiation relative to the excitation wavelength, the fluorescence emission is also detected as re-emitted (not absorbed) radiation. Therefore, the remitted light relative to the fluorescence should be subtracted from the spectrum. Nevertheless, it is not possible to distinguish its contribution because the Lambda 900 spectrometer is provided only with one monochromator, the one that selects the excitation wavelength, but no emission monochromator is present. This means that for every excitation wavelength all the emitted wavelengths are detected, both the reflected and the re-emitted ones as luminescence emission. Without emission monochromator it is not possible to distinguish between reflected light and luminescence emission.

Therefore, reflection measurements were performed on the Cary Eclipse Varian spectrophotometer. Reflection in the place of remission indicates the detection of the reflected light with the same wavelength as the excitation light, without the contribution of all re-emitted beams. The Varian Eclipse spectrometer contains an emission monochromator, as well as the excitation monochromator. In this instrument, it is possible to synchronize the monochromators in order to detect only the wavelength relative to the exciting laser. In this way it is possible to state, for every wavelength, the amount of light reflected, without considering other wavelengths generated by the luminescence mechanism. Therefore, more accurate results can be obtained for luminescent samples. The disadvantage is the lack of an integrating sphere, which detects the reflected light in all directions.

5.4.2.1. Remission measurements

A Perkin Elmer Lambda 900 UV/VIS/NIR spectrometer was used to carry out the optical measurements, using an integrating sphere. $BaSO_4$ was used as white reference for the measurement of MK600; afterwards MK600 was used as the reference material for all the remaining samples. Since $BaSO_4$ is perfectly reflecting only in the range 335-1320 nm, the spectra are not reliable for more energetic wavelengths. The normalized remission spectrum on the reference was converted in an absorption/scattering spectrum using the Kubelka-Munk function. The samples were measured in powder form in 1 mm precision cells made of Quartz Suprasil®, 100-QS, Hellma®. All measurements were performed in air at room temperature. Remission measurements were initially performed on the samples MK, Ceraset and KiON S. Afterwards, also reflection measurements were carried out on these samples, besides the remission measurements. For the remaining samples, only reflection measurements were performed.

5.4.2.2. Reflection measurements

The reflection measurements were recorded on the Cary Eclipse Varian spectrophotometer, using a Xenon lamp as the excitation source. The measurements were performed in mode "Synchronized", meaning that excitation and emission monochromators were synchronized on the same wavelength, i.e. only the exciting wavelength is detected in the emission monochromator. The simultaneous detection allowed the record of a reflection spectrum. The original spectrum was normalized on a $BaSO_4$ white reference (perfectly reflecting only in the range 335-1320 nm). The normalized reflection spectrum was subsequently converted in an absorption/scattering spectrum using the Kubelka-Munk function. The samples were measured in powder form in 1 mm precision cells made of Quartz Suprasil®, 100-QS, Hellma®. The cell was positioned at 35° of angle in order to minimize the scattering. All measurements were performed in air at room temperature.

5.5. Structural characterizations

Vibrational spectroscopy (FT-IR and Raman) and solid state MAS NMR (^{29}Si and ^{13}C) are the main structural tools applied to investigate the polymer-to-ceramic transformation. The goal of these characterizations is a better understanding of the processes that take place during annealing of the polymers. FT-IR and Raman spectra provide information about the chemical bonds present in the studied materials and, when heat-treated samples are measured, the bond rearrangement with temperature. ^{29}Si and ^{13}C MAS NMR measurements identify the different Si and C environments and their changes during heat-treatment. With ^{29}Si MAS NMR it is possible to determine the polymer-ceramic transformation by analyzing the elements bonded to the Si. ^{13}C MAS NMR offers an analysis of the changes on the C sites and detects the formation of different carbon species such as aromatic sp^2 carbon [Ionescu2009a].

5.5.1. FT-IR spectroscopy

Infrared spectra (FT-IR, Fourier Transform Infra Red) were recorded on a Perkin Elmer 1750 Infrared Fourier Transform spectrometer, either using KBr discs with a spectral resolution of 2.0 cm^{-1} or with a single reflection ATR system (MKII Golden Gate^{TM}, Specac) and spectral resolution of 1.0 cm^{-1}, in both cases with spectral range from 4000 to 400 cm^{-1}. Since the ATR method is not reliable for wavenumbers smaller than 600 cm^{-1}, only the reliable range was considered and shown in the graphs. When needed, the spectra were normalized in order to illustrate comparable transmittance intensities on the same graph. All measurements were performed in air (in argon for polysilylcarbodiimides) at room temperature.

5.5.2. Raman spectroscopy

Raman spectra were recorded on a confocal Horiba Jobin Yvon HR 800 micro-Raman spectrometer with excitation laser wavelengths of 514, 488 and 633 nm. To avoid interferences with the fluorescence background, a IR/Raman spectrometer Bruker IFS 55 - FRA 106, using a laser ND:YAG of wavelength 1064 nm was also utilized. All measurements were performed in air at room temperature.

5.5.3. NMR spectroscopy

5.5.3.1. Liquid State NMR

The ^{1}H, ^{13}C and ^{29}Si liquid state NMR (Nuclear Magnetic Resonance) measurements were carried out on a 500 MHz-Spectrometer Bruker DRX 500 (AVANCE 500) using a 5 mm BBO rotor diameter. All spectra were referenced to tetramethylsilane (TMS) as the internal standard (reference), which corresponds to chemical shift δ = 0 ppm. The measurements were carried out at frequencies of 500 MHz (^{1}H), 126 MHz (^{13}C) and 99 MHz (^{29}Si). All samples were measured dissolved in deuterated benzene (C_6D_6).

5.5.3.2. Solid State NMR

Solid state Magic Angle Spinning NMR (MAS NMR) as well as Cross Polarization coupled with Magic Angle Spinning (CP MAS NMR) experiments were recorded with a Bruker AVANCE 300 spectrometer using a 4 mm rotor diameter. ^{29}Si and ^{13}C NMR experiments of MK and Ceraset samples were carried out at frequencies of 59.62 MHz and 75.47 MHz respectively. VL20 and GV samples were measured at frequencies of 59.63 (^{29}Si), 75.48 MHz (^{13}C) and 300.29 MHz (^{1}H) using 4 mm rotors spinning at 10 kHz for ^{13}C (except for GV treated at 300, 400 and 500 °C recorded at 5 kHz) and ^{29}Si and 14 kHz for ^{1}H. For ^{29}Si MAS experiments 45° pulses (2.8 µs) and 60 s recycle delays were used, to account for the long T_1 relaxation time of ^{29}Si nuclei in such samples. Chemical shifts are referred to tetramethylsilane (TMS). The solid state NMR spectra were simulated using a specific program (dmfit2008).

5.6. Thermal analysis

Thermal Gravimetric Analysis (TGA) coupled with Mass Spectrometry (MS) (TGA/MS) is also an important tool to investigate the polymer-to-ceramic transformation. TGA/MS detects the mass losses and the gaseous species released during the conversion from the polymeric to the ceramic state [Ionescu2009]. The TGA/DTA/MS measurements (DTA being Differential Thermal Analysis), or STA (Simultaneous Thermal Analysis), were carried out using STA 449C Jupiter coupled with QMS 403C Aëolos, Netzsch Gerätebau GmbH, Selb/Bayern. The samples were heated to 1400 °C with a heating rate of 5 °C min^{-1} in argon atmosphere, while simultaneously measuring the mass loss and the gases released via mass spectrometry (Quadrupole Mass Spectrometer). DTG (Differential Thermal

Gravimetric Analysis) curves were obtained by differentiation of the TGA curves. For some sample also *in situ* FT-IR of the gaseous byproducts was acquired.

5.7. X-Ray Diffraction (XRD)

The X-ray data were collected on a STADI P diffractometer from the company STOE (Cu $K_{\alpha 1}$ radiation) provided with a PSD (Position Sensitive Detector). The structures were refined using the program STOE WinXPOW.

5.8. EPR spectroscopy

Electron paramagnetic resonance (EPR) or electron spin resonance (ESR) spectroscopy is a technique for studying chemical species that have one or more unpaired electrons, for example free radicals.

The principles of EPR analysis are here briefly presented. When an unpaired electron is located in an electromagnetic field B, it aligns to it either in a parallel or anti-parallel direction. The parallel condition is at lower energy than the anti-parallel condition. If an amount of energy equal to the energy difference between the two states is absorbed by the unpaired electron from an external source, the unpaired electron will be in resonance between the two states (Figure 2).

5. Experimental procedure

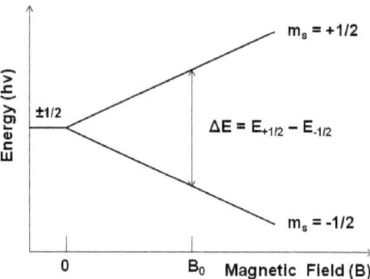

Figure 2: Resonance condition of the spin between parallel and anti-parallel conditions [Vij2006].

In the experimental practice, the energy from the external source (microwaves) is kept constant and the magnetic field is varied. When the magnetic field value is such to satisfy the resonance equation, the unpaired electron absorbs the microwaves energy and reaches the resonance condition. A signal is revealed in the EPR spectrum. Thus, EPR is based mainly on an equation, the resonance condition:

$$h\upsilon = \beta g B.$$

In EPR measurements the energy is provided by a constant microwaves source with frequency υ=9.31 GHz. The Planck constant is equal to $6.63 \cdot 10^{-34}$ J·s. β is the Bohr magneton constant ($9.27 \cdot 10^{-24}$ J·T^{-1}). The magnetic field B is varied and measured in the resonance condition. Its value is around 330 mT. All the terms of the resonance equation are known; therefore the g-factor can be acquired in the resonance condition, considering the value of B. The electron g-factor, also called Landé factor, is a fingerprint of the radical. Consequently, it denotes the radical species present in the material [Wertz1972, Vij2006].

In the EPR absorption curve, which represents the energy absorbed by the unpaired electron, the Full Width at Half Maximum (FWHM), called line width, is useful to detect the hybridization state of the unpaired electron [Prasad2000, Trassl2002].

The integrated area of the measured curve (which is the derivative of the absorption) denotes the intensity of the signal, which is proportional to the spin concentration. Using a suitable reference (in our case polycrystalline DPPH (2,2-Diphenyl-1-Picrylhydrazyl)) and considering the sample weight, the spin concentration can be extrapolated [Barklie2000].

The X band (9.5 GHz) continuous wave EPR measurements were performed using an ESP 300E spectrometer (Bruker) equipped with a rectangular TE_{112} resonator. The magnetic field was read out with a NMR gauss meter (ER 035M, Bruker). All spectra were recorded at 15 K using an Oxford cryostat. Additionally, a standard field marker (polycrystalline DPPH with g=2.0036) was used for the calibration of the resonance magnetic field values and the determination of the exact g-factor of the samples. The EPR spectra were best fitted with a superposition of one or two Lorentzian absorption lines in MathWorks, Inc. software.

5.9. Note

None of the samples investigated showed phosphorescence emission; therefore no spectra will be reported. Furthermore, in lifetime measurements no results were obtained due to the too short lifetimes of the polymers compared to the lamp emission decay. Therefore, also in this case no data will be reported. Considering the phosphorescence and lifetime results, we can attest that the photoluminescence mechanism of the heat-treated silicon-based samples is the fluorescence (electric-dipole allowed transitions).

5.10. References

[Barklie2000] R. C. Barklie, M. Collins, S. R. P. Silva, *Phys. Rev. B*, 61[5], 2000

[CIE] CIE *Commission Internationale de l'Eclairage Proceedings, 1931*, Cambridge University Press, Cambridge, 1932

[Kortüm1969] G. Kortüm, *Reflectance Spectroscopy: Principles, Methods, Applications*, Springer, 1969

[Ionescu2009a] E. Ionescu, C. Gervais, F. Babonneau, *The Polymer-to-Ceramic Transformation*, Ch 2.3 in: P. Colombo, G.D. Sorarù, R. Riedel, A. Kleebe (Eds.), *Polymer-derived-ceramics*, DESTech Publications, Lancaster, PA, 2009.

[Mera2009b] G. Mera, R. Riedel, F. Poli, K. Müller, *J. Eur. Ceram. Soc.*, 29, *2873-2883*, 2009

[Prasad2000] B. L. Prasad, H. Sato, T. Enoki, Y. Hishiyam, Y. Kaburagi, A. M. Rao, P. C. Eklund, K. Oshida, M. Endo, *Phys. Rev. B*, 62, *11209-11218*, 2000

[Trassl2002] S. Trassl, G. Motz, E. Rössler, G. Ziegler *J. Am. Ceram. Soc.*, 85 [1], *239*, 2002

[Vij2006] D. R. Vij, *Applied Solid State Spectroscopy*, Springer, 2006

[Wendlandt1966] W. WM. Wendlandt, H. G. Hecht, *Reflectance Spectroscopy*, Interscience Publishers/John Wiley, 1966

[Wertz1972] J. E. Wertz, J. R. Bolton, *Electron spin resonance: Elementary Theory and Practical Applications,* New York, McGraw-Hill Book Company, 1972

6. Results and discussion

6.1. Polysiloxanes

6.1.1. Wacker-Belsil® PMS MK polymethylsilsesquioxane

Polymethylsilsesquioxane Wacker-Belsil®PMS MK (MK polymer) (Wacker Chemie GmbH, Burghausen), was chosen because of its commercial availability, non-toxicity, low cost, mouldability and crosslinkability. Wacker-Belsil®PMS MK is a solid, solvent-free and highly crosslinked silicone resin. It is odorless and colorless, its softening point falls between 50 and 60 °C and it is soluble in organic solvents (aromatic solvents and ketones) [Wacker, Harsche2004b].

The structure of MK polymer is not provided by Wacker and thus partially still unknown. Segments of the structure were proposed by Harsche on the basis that the polymer is highly crosslinked and possesses, as functional units, approximately 2 mol % of hydroxy and ethoxy groups [Harsche2004b]. The structure segments, which are in accordance with FT-IR observations, are illustrated in Figure 3.

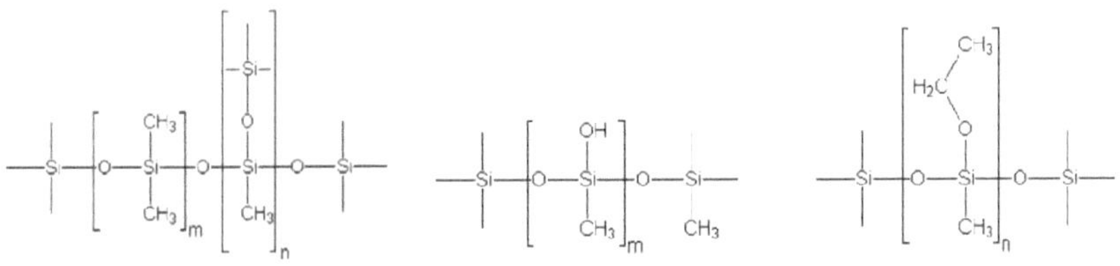

Figure 3: Structure segments of MK polymer with linear and branched components and with functional groups (hydroxy and ethoxy).

6. Results and discussion

It is clear that MK can crosslink through hydroxy and ethoxy groups at lower annealing temperatures and through methyl groups at higher temperatures. Recently, the release of cyclic structures during annealing, and consequently their presence in the starting polymer, was attested by STA measurements, thanks to *in situ* FT-IR measurements. In Figure 4, the presence of R groups in the MK structure is reported, where R represents the cyclic structure octamethylsilsesquioxane [Ionescu2009b].

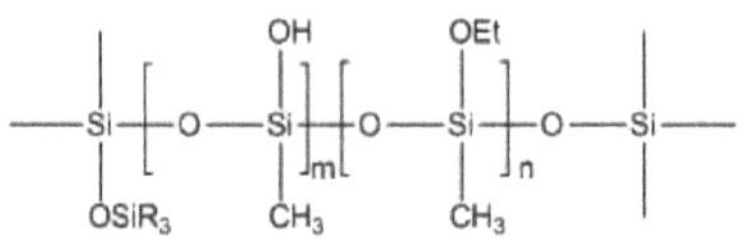

$$-\text{Si}\left[-\text{O}-\underset{\underset{\text{OSiR}_3}{|}}{\text{Si}}\right]-\text{O}-\underset{\underset{\text{CH}_3}{|}}{\overset{\overset{\text{OH}}{|}}{\text{Si}}}\Bigg]_m\left[-\text{O}-\underset{\underset{\text{CH}_3}{|}}{\overset{\overset{\text{OEt}}{|}}{\text{Si}}}\right]_n-\text{O}-\text{Si}-$$

Figure 4: Structure of MK polymer with presence of octamethylsilsesquioxane groups (R) [Ionescu2009b].

The chemical composition of MK proposed by Ionescu *et al.* is $SiO_3C_1H_{3.3}$ [Ionescu2009b]. The MK polymer is commonly used in cosmetics and as precursor for SiOC ceramics [Wacker]. In the last case it can be used for several applications, such as structural ceramic for high-temperature applications, ceramic foams, micro-electro-mechanical-systems (MEMS), metal matrix composites (MMC) and ceramic matrix composites (CMC) [Ionescu2009b, Balan2006, Ischenko2006, Schneider1988, Takahashi 2001, Wilhelm2005, Anggono2006, Bernardo2006, Bernardo2007, Bernardo2009, Colombo2002, Colombo2003, Maire2007, Colombo2008, Costacurta2007, Friedel2005, Harshe2004a, Herzog2005, Kim2007, Ritzhaupt-Kleissl2006, Moysan2007, Nangrejo2009, Rambo2006a, Rambo2006b, Rocha2005, Rocha2006, Rocha2008, Scheffler2002, Scheffler2003, Takahashi 2003, Thünemann2007, Vaucher2008, Wei2002].

In this work, the MK polymer was heat-treated at 200, 300, 400, 500, 600, and 700 °C for 2 h under Ar flow, with heating rate 50 °C/h and free cooling. The MK powder melted and crosslinked, and single piece samples with the shape of the mold were obtained, which were subsequently powdered.

6.1.1.1. Photoluminescence measurements

6.1.1.1.1. Fluorescence measurements

Surprisingly, photoluminescence properties were observed for all samples [Menapace2008]. Figure 5 shows the fluorescence emission and excitation spectra of the MK samples relative to excitation and emission wavelengths according to Table 1, corresponding to the maximum emission and excitation peaks. The spectra were obtained within a week after the thermal treatment.

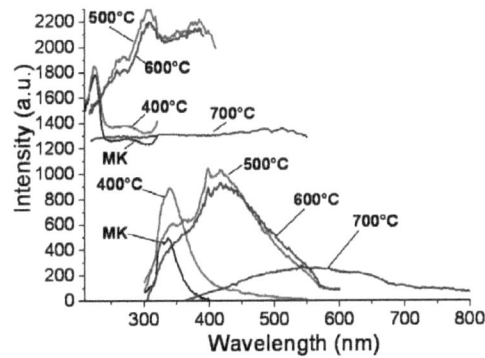

Figure 5: Fluorescence emission (bottom right) and excitation (top left) spectra of the MK samples heat-treated at different temperatures (labeled). The excitation and emission wavelengths applied are described in Table 1.

Table 1: Excitation and emission wavelengths used to obtain fluorescence spectra for the MK samples. They correspond to the maximum emission and excitation peaks.

Sample	Excitation wavelength (nm)	Emission wavelength (nm)
MK polymer	225	336
MK 200 °C	225	336
MK 300 °C	225	336
MK 400 °C	230	336
MK 500 °C	306	434
MK 600 °C	306	434
MK 700 °C	350	565

The MK polymer and the samples treated at low temperatures (200 and 300 °C) show photoluminescence in the UV range and are characterized by analogous spectra. Therefore, for clarity reasons, only the emission and excitation spectra of MK polymer are illustrated in Figure 5.

The main excitation peak is at around 225 nm and there are some weak excitation peaks at lower energies. The main emission peak is at 336 nm but it is possible to individuate the contribution of other peaks. The same bands are present in the three samples annealed up to 300 °C with slightly different relative intensities.

After heat-treatment at 400 °C, an increased contribution of the low energy bands was observed in both the excitation and emission spectra. The main excitation peak is at 230 nm, while the main emission peak is still at 336 nm. The samples heat-treated up to 400 °C are transparent and colorless and show fluorescence in the UV range. These properties could be useful for UV-LEDs.

The samples heat-treated at 500 and 600 °C reveal the maximum relative values of fluorescence intensity. Moreover, the fluorescence emission is in the visible range. The materials are translucent and possess a yellow-brown color. Besides the excitation peaks present in the previous samples, new bands appear at lower energies. Also in the emission spectra new red-shifted peaks are detectable at 366 nm, 403 nm and 443 nm. A part of the emission spectrum is still in the UV range, but the newly formed peaks in the visible range are responsible for the visible photoluminescence of the samples.

Following treatment at 700 °C, the luminescence intensity drastically decreases, because of the formation of free carbon, as attested from MAS NMR measurements (shown later), which renders the material dark and opaque. The maximum excitation and emission spectra are red-shifted and their intensities are sensitively decreased.

From the viewpoint of LED applications it is interesting to investigate the photoluminescence properties obtained with excitation wavelength in the range 360-400 nm [Rohwer2003, Green1997]. The emission spectra obtained with 360 nm as the excitation wavelength of the samples MK 400-700 °C are represented in Figure 6. Samples annealed at lower temperatures did not show fluorescence when excited with 360 nm.

6. Results and discussion

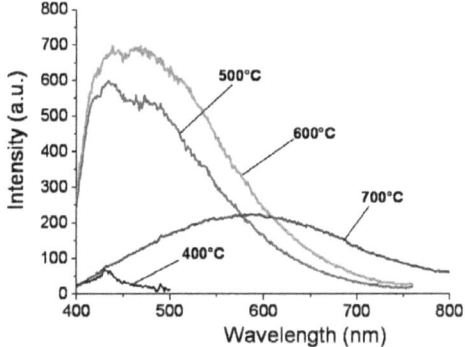

Figure 6: Fluorescence emission spectra of the MK samples heat-treated at different temperatures (400-700 °C) with excitation wavelength 360 nm.

The intensities of the emission spectra relative to MK 500 and 600 °C obtained using 360 nm as excitation wavelength are lower than those obtained with more energetic excitation wavelengths. Furthermore, the spectra are shifted to lower energies. Correspondingly, only the range at lower energies of the maximum excitation spectra is excited and only the visible part of the entire emission spectra is emitted.

The fluorescence spectra of the MK samples obtained with excitation wavelength 250 nm (Figure 10 (a)) display similar features as the maximum emission intensity spectra.

In Figure 7, powdered MK samples annealed at different temperatures (400, 500, 600 and 700 °C) are shown under white (a) and UV (360-400 nm) (b) light.

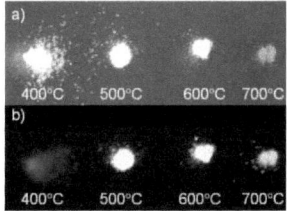

Figure 7: Photographs representing MK samples annealed at different temperatures in powder form, under white (a) and UV (360-400 nm) (b) light, respectively. For colored pictures the reader can refer to http://tuprints.ulb.tu-darmstadt.de/2085/.

6. Results and discussion

The white photoluminescence of the samples heat-treated at 500 and 600 °C is remarkable. In Figure 8, MK annealed in a temperature gradient between 650 and 700 °C is shown under white (a) and UV (360-400 nm) (b) light.

Figure 8: Photographs representing MK annealed in the presence of a temperature gradient between 650 and 700 °C, under white (a) and UV (360-400 nm) (b) light, respectively. For colored pictures the reader can refer to http://tuprints.ulb.tu-darmstadt.de/2085/.

The material annealed at 650 °C, like the samples heat-treated at 500 and 600 °C, is translucent and brown colored under white light and shows white photoluminescence and opacity under UV irradiation. After annealing at 700 °C the fluorescence is quenched and the sample appears black. Small temperature differences in the annealing temperature have significant effects on the optical and photoluminescence properties of the samples.

The spectral emissions of luminescent samples can be related to color coordinates in the CIE chromaticity diagram, where the emission colors correspond to a point in the graph. The spectral emissions obtained with 360 nm excitation of the MK samples heat-treated at 400-700 °C are illustrated on the CIE chromaticity diagram 1931 (Figure 9).

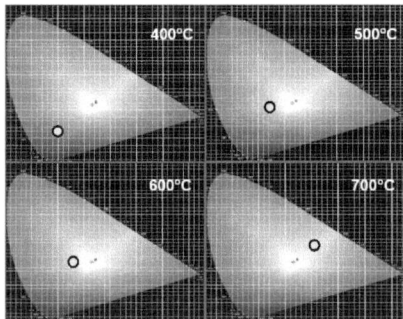

Figure 9: CIE chromaticity diagrams 1931 of the MK samples annealed at 400-700 °C. For colored pictures the reader can refer to http://tuprints.ulb.tu-darmstadt.de/2085/.

The color coordinates are in accordance with the colors of the samples under UV light (Figure 7).

6.1.1.1.2. Effect of the annealing atmosphere

In order to observe the effect of the annealing atmosphere on the fluorescence properties of the heat-treated polymers, a set of MK samples was also annealed in a reducing atmosphere of Ar/H_2 gas mixture (93.5% Ar, 6.5% H_2).
The emission spectra of the polysiloxane annealed in argon (a) and in reducing atmosphere of Ar/H_2 (b) at different temperatures, both obtained with excitation wavelength of 250 nm, are reported below (Figure 10).

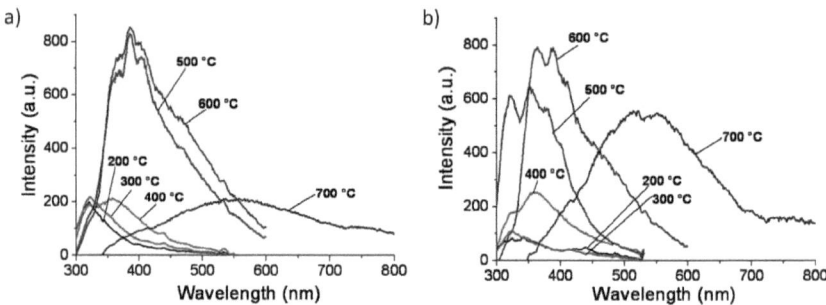

Figure 10: Fluorescence emission spectra of MK samples annealed in argon (a) and in reducing atmosphere of Ar/H_2 (b) at different temperatures, both obtained with excitation wavelength 250 nm.

The annealing atmosphere (Ar or Ar/H_2) seems to play a negligible role on the photoluminescence properties of MK up to 400 °C. The luminescence intensity of the samples treated in Ar/H_2 is inferior up to 300 °C, but the main peaks have roughly the same shape at the same temperatures. The spectra relative to the samples treated at 400 °C display similar intensities.

After annealing of MK at 500 °C, the emission spectra obtained in the two different atmospheres are different: the spectrum relative to the sample annealed in Ar/H_2 is blue-shifted and less intense compared to the sample annealed in Ar. During heat-treatment in Ar, between 400 and 500 °C, structural rearrangements lead to the formation of new red-shifted peaks in the emission spectrum. These peaks are not yet developed at that temperature under Ar/H_2 atmosphere. The peak at 336 nm, which is only a shoulder for the MK sample heat-treated at 500 °C in Ar, is among the most intense peaks in Ar/H_2, as for the samples annealed up to 400 °C. Considering these observations, it can be attested that the reducing atmosphere retards the appearance of chemical species which produce the visible luminescence.

At 600 °C, similar emission spectra are obtained for the different atmospheres, although the spectrum obtained in Ar/H_2 is more similar to that of MK 500 °C than that of MK 600 °C treated under Ar. This means that the appearance of species responsible for the visible luminescence, occurring in Ar between 400 and 500 °C, is shifted between 500 and 600 °C in Ar/H_2.

After annealing at 700 °C in Ar/H_2, the material is characterized by a stronger luminescence intensity compared to that of the sample annealed at the same temperature in Ar. This can be related to a smaller amount of free carbon phase developed, responsible for the reduction of the emission intensity.

Liemersdorf et al. demonstrated that the presence of hydrogen in the annealing atmosphere causes carbon loss from the sample [Liemersdorf2008]. This could be a reason of the reduced presence of carbon in the sample after annealing at 700 °C.

Thus, in the presence of a reducing atmosphere, the same luminescence properties seem to develop at higher temperatures in comparison to annealing in argon.

6.1.1.1.3. Stability of the fluorescence during storage

The MK samples were stored in air and the luminescence properties were measured several times. In Figure 11, the emission spectra of the MK samples measured within the first week after heat-treatment and two years later, obtained with excitation wavelength 250 nm, are shown.

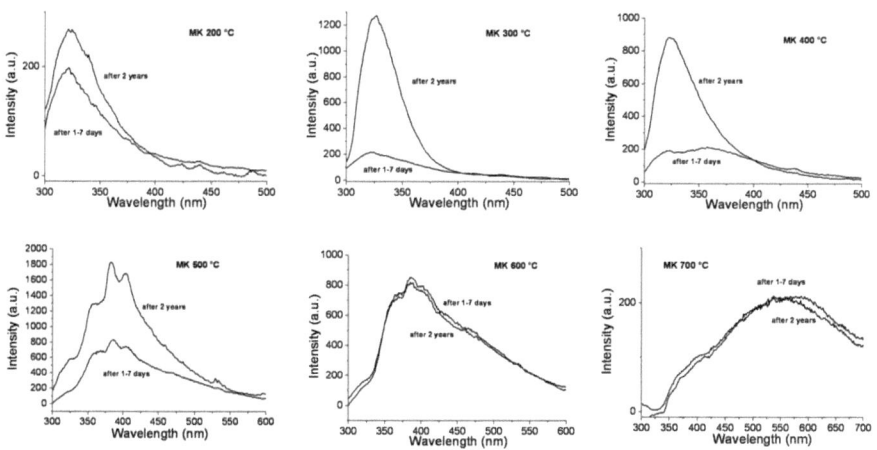

Figure 11: Fluorescence spectra of MK samples annealed in argon at different temperatures, measured during the first week and 2 years after the heat-treatment, obtained with excitation wavelength 250 nm.

After heat-treatment up to 500 °C, the powdered MK samples show an increase in the emission intensity after two years of storage in air. Although the polysiloxanes are air stable, the annealed samples are not completely insensitive to air. Probably, functional groups that can react with moisture, for example hydroxy groups, are still present in the polymeric samples. Moreover, the aging of the samples could result in enhanced crosslinking. In all cases the contact with air has a positive effect on the luminescence emission. On the contrary, the samples heat-treated at higher temperatures (600 and 700 °C) show unchanged emission and thus their photoluminescence properties are not affected by storage in air. The MK samples annealed at 400-700 °C were also studied 5 months after the heat-treatment (not shown here). Except for some slight intensity changes, which are within the measurement error, no variation in the emission bands and their ratio can be observed. Therefore, in the first months after heat-treatment the fluorescence emission changes only slightly, while the major changes occur in the following year.

6.1.1.1.4. Quantum efficiency

The quantum efficiency was not measured on any of the MK samples. The quantum efficiencies of the MK samples were estimated by comparing the integral of their emission spectra (in energy scale) with the integral of the emission spectra of samples with known quantum efficiency, having emission in the same range (Table 2).

Table 2: Quantum efficiencies of the MK samples.

Sample	Quantum efficiency (%)
MK polymer	4.5*
MK 200 °C	3.2*
MK 300 °C	3.9*
MK 400 °C	11.4*
MK 500 °C	13.4
MK 600 °C	11.8
MK 700 °C	2.3

The samples designed with * emit in the UV range, for which no measurement was successful. Therefore the values were calculated on comparison with a sample with emission maximum at 394 nm. The values obtained are quite high if compared with quantum efficiencies of similar materials, like doped glasses (q.e. = 8-60%) [Andrade2006, Suzuki2005, Quimby1998], or newly developed light emitting materials, for example silicon and silicon nanocrystals (q.e. = 0.01-1%, 19%) [Green2001, Valenta2004]. In addition, the values related to the MK samples were obtained with a simple route, which is still to be improved.

6.1.1.2. Absorption measurements

Additional information about the electronic transitions occurring in the MK samples and their variation after temperature treatment can be obtained with absorption measurements.

6.1.1.2.1. UV-Vis-NIR spectroscopy

Standard UV-Vis measurements were performed in THF as solvent. Nevertheless, the solubility of MK decreases as the treatment temperature increases, along with the transformation from the polymer to the polymer-ceramic state. MK was still soluble after heat-treatment up to 200 °C, therefore reasonable absorption spectra were obtained only for the MK polymer and the sample annealed at 200 °C, in THF solution (Figure 12).

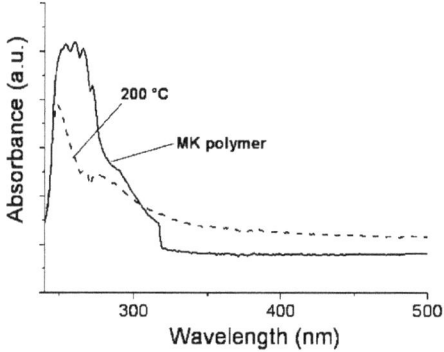

Figure 12: UV-Vis absorption spectra of the MK polymer (solid line) and MK annealed at 200 °C (dashed line).

The UV-Vis analysis shows that the MK polymer and MK after annealing at 200 °C absorb in the UV range. The absorption edges occur for both samples at around 320 nm (3.87 eV). The step at 319 nm is due to the lamp switch.

6.1.1.2.2. Remission measurements

Since absorption measurements in solution are not suitable for heat-treated polymers, remission measurements were performed on powdered MK samples, using the same spectrometer (Perkin Elmer Lambda 900) and utilizing an Ulbricht (integrating, photometer) sphere. $BaSO_4$ was used as white reference for the measurement of MK600; afterwards MK600 was used as the reference material for all the remaining samples. Absorption/scattering spectra were obtained by elaboration of the remission data, using the Kubelka-Munk function. Although it is not so accurate for luminescent materials, as explained in the experimental procedure section, the remission spectra give an idea of the absorption range of the samples (Figure 13).

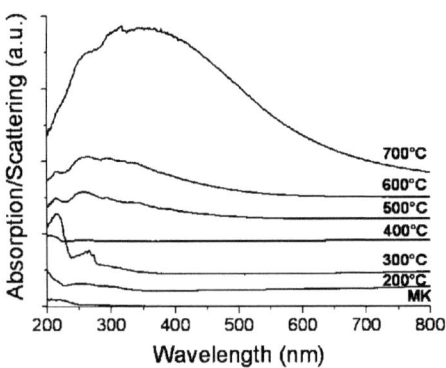

Figure 13: Absorption/scattering spectra of the MK samples obtained with remission measurements.

The samples annealed up to 400 °C absorb in the UV range, below 350 nm. For heat-treatments from 500 °C the samples also absorb in the visible range. With increasing treatment temperature from 500 to 600 °C, only a slight red-shift of the absorption edges is observable, while from 600 to 700 °C the red-shift is significant, exactly as in the photoluminescence spectra. Since absorption in the visible range implies coloration of the material, from 500 °C coloration of the samples is expected. This is in accordance with the optical observations (see pictures in Figure 7 and 8), where the samples MK 500 and 600 °C show a slightly yellow coloration, while MK 700 °C is brown-black, due to its absorption in the whole visible range.

6.1.1.2.3. Reflection measurements

More reliable absorption/scattering spectra were obtained using synchronous excitation and emission monochromators, on the Cary Eclipse Varian spectrophotometer (reflection measurements). $BaSO_4$ was used as a white reference material.
The absorption/scattering spectra (Kubelka-Munk function) obtained from the reflection data relative to the MK samples, are presented in Figure 14.

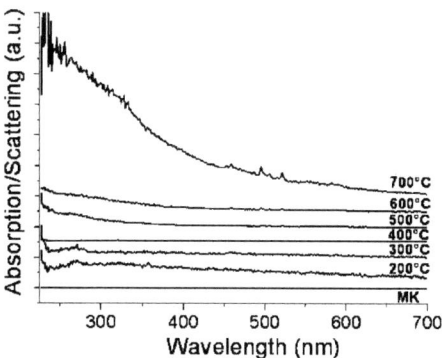

Figure 14: Absorption/scattering spectra relative to the MK samples obtained with reflection measurements.

The samples annealed up to 400 °C do not show absorption in the measured range. This is in contrast with the UV-Vis measurements of MK and MK 200 °C, which absorb in the UV range up to 320 nm. However, it must be taken into account that the reflection measurements are reliable only in the range 335-1320 nm, where the white reference $BaSO_4$ is perfectly reflecting. The samples annealed at 500 and 600 °C show absorption edges at about 400 nm (3.10 eV), while MK 700 °C shows absorption in the whole detected spectral range up to 700 nm (1.77 eV). The results are in accordance with the excitation spectra of Figure 5, with the remission measurements and with the optical observations. The absorption spectra show trends similar to the photoluminescence spectra, as the absorption edge is red-shifted as the treatment temperature increases, exactly as the maximum excitation and emission spectra from photoluminescence measurements. The red-shifts are appreciable from 400 to 500 °C and from 600 to 700 °C. Moreover, up to 400 °C the absorption, excitation and emission spectra are in the UV range, while from 500 °C the spectra are shifted toward the visible range.

6.1.1.3. FT-IR spectroscopy

The MK polymer and the heat-treated MK samples were investigated by means of FT-IR structural analysis (Figure 15). In Table 3, the vibration bands and respective bonds are listed. The measurements were obtained on discs of KBr mixed with the samples, except MK 200 and 300 °C, for which the ATR device was used.

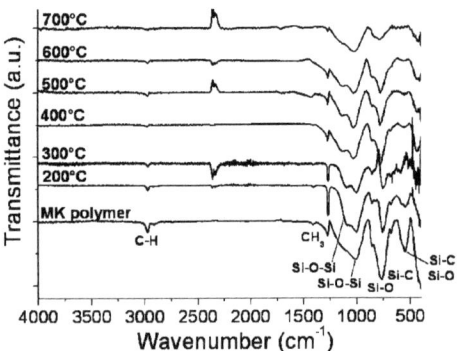

Figure 15: FT-IR spectra of the MK samples.

Table 3: Vibration bands and respective bonds detected by means of FT-IR analysis of the MK samples.

Bond assignment	Vibration band (cm^{-1})
$v_{as}(-CH_3)$	2976
δ C-H (Si-CH$_3$)	1276
$v_{as}(Si-O-Si)TO$	1190
$v_a(Si-O-Si)LO$	1090
$v_a(Si-O-Si)6r$	1040
$D(Q)(SiO_2C_2)$	860
$v(Si-O-C)(ceramic)$	812
$v_a(Si-O)$	786
$v_a(Si-C)$	768
$v(Si-O)$	576
$v(Si-C)$	548

The FT-IR analysis does not highlight any relevant transformations up to 600 °C, probably because the MK polymer is a highly crosslinked polysiloxane in the precursor state. The disappearance of Si-CH$_3$ bonds (1276 cm^{-1} and 768 cm^{-1}) at 700 °C denotes the cleavage of the Si-CH$_3$ bonds, with elimination of methane and hydrogen, and their replacement with Si-O-Si and Si-O-C bonds (812 cm^{-1}), which indicate the formation of a ceramic network. The band at 576 cm^{-1} (Si-O), due to Si-O-Si in rings, is remarkably reduced after heat-treatment at 200 °C and disappears at 700 °C. The intensity reduction after annealing at 200 °C suggests that some of the cyclic structures evolute at this temperature (as later shown in STA analysis) [Ionescu2009b]. Another reason of the band decrease is the ring opening which enhances the crosslinking effect. It is well known that this band is displaced at lower frequencies in more open structures.

6.1.1.4. Raman spectroscopy

The heat-treated MK samples were investigated with confocal micro-Raman spectrometer (Figure 16 and Table 4). An excitation laser of 633 nm was found to be optimal in order to minimize the fluorescence background.

6. Results and discussion

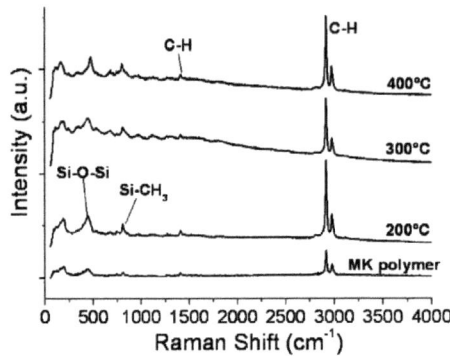

Figure 16: Raman spectra of the MK samples using 633 nm as laser wavelength.

Table 4: Vibration bands and respective bonds detected by means of Raman analysis of the MK samples.

Bond assignment	Raman shift (cm^{-1})
$v_{as}(-CH_3)$	2974
$v_{as}(CH_2)$	2914
$\delta_{as}(C-H)$	1410
$\delta_s(C-H)\ CH_3$	1267
$v_s(Si-OH)$	951
$\rho(-Si(CH_3))$	803
$v_s(C-Si-O)$	768
$v_s(C-Si-C)$	741
CH_3 rocking	679
$\delta(O-Si-O)$	440-480
$\delta(C-Si-O)$	236
$\delta(C-Si-C)$	190
C-Si-C twist	157

In the MK samples annealed up to 400 °C common Raman bands can be individuated. The bands denote the presence of CH_3 (asymmetrical vibration) or CH_2 bonds (1410 cm^{-1}), CH_2 bonds (2914 cm^{-1}), -CH_3 bonds (2974, 1267, 679 cm^{-1}), Si-OH bonds (951 cm^{-1}), -$SiCH_3$ bonds (803 cm^{-1}), C-Si-O bonds (768, 236 cm^{-1}), C-Si-C bonds (741, 190, 157 cm^{-1}) and Si-O-Si bonds (440-480 cm^{-1}). In the samples annealed from 500 °C the fluorescence background interferes with the measurements and any analysis is hindered. For this reason, the samples heat-treated at 400-700 °C were also analyzed with the IR/Raman spectrometer Bruker IFS 55 - FRA 106, with laser wavelength 1064 nm. The spectra are reported in Figure 17.

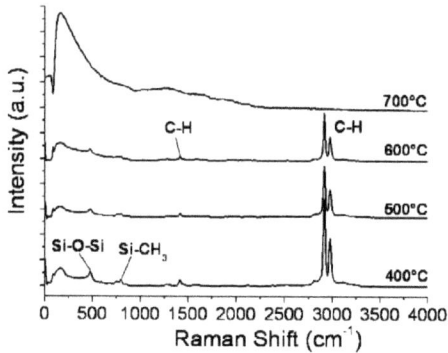

Figure 17: Raman spectra of the MK samples heat-treated at 400-700 °C using 1064 nm as laser wavelength.

With 1064 nm as the laser wavelength the luminescence interference is avoided. The CH_2 bonds (2914 cm^{-1}) and -CH_3 bonds (2974, 1259 cm^{-1}), the CH_3 (asymmetrical vibration) or CH_2 bonds (1410 cm^{-1}), as well as the Si-O-Si (474 cm^{-1}), -$SiCH_3$ (793-802 cm^{-1}) and C-Si-C (741, 160 cm^{-1}) bonds are clearly detectable up to 600 °C. The sample heat-treated at 700 °C shows some interference, which is not attributed to luminescence background but to the heating of the carbon clusters. Therefore for this sample the Raman analysis is hindered.

6.1.1.5. TGA/DTG/MS

Thermal analysis was carried out on the MK precursor, in order to investigate the thermal decompositions and transformations occurring during annealing. The TGA- and its differential DTG-curves of polysiloxane MK, obtained in argon, are illustrated in Figure 18. The thermal transformation of MK in nitrogen and air was studied in literature [Schneider1988, Kim2007, Rocha2008,

Takahashi2003, Anggono2006], while the decomposition in argon atmosphere was shown in the dissertation of Harsche [Harsche2004b] and recently reported by Ionescu et al. [Ionescu2009b].

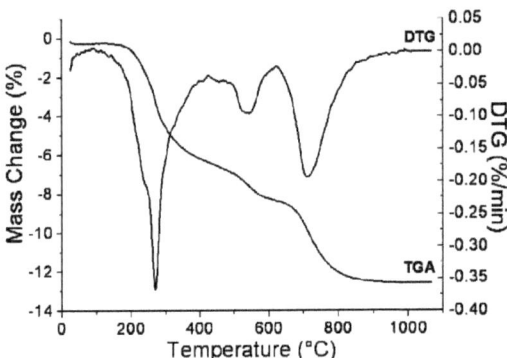

Figure 18: TGA and DTG of MK polymer in argon.

In Figure 19, selected MS spectra of gaseous byproducts relative to the decomposition of MK polymer are shown.

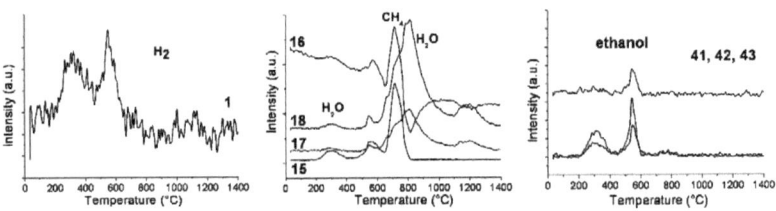

Figure 19: Selected TG/MS analysis of the gaseous byproducts of the thermal decomposition of the MK polymer. The numbers in the graphs refer to m/z.

The STA graphs display three main mass losses. The first step up to 400 °C is due to the evaporation of low molecular weight oligosiloxanes, to hydrogen elimination and to polycondensation reactions involving the loss of H_2O and ethanol. In the second step, between 500 and 600 °C, oligosiloxanes and ethanol evolve, and mineralization reactions start to take place with release of H_2 and CH_4. The third step, between 600 and 800 °C, is mainly associated with water loss and ceramization reactions resulting in the loss of H_2 and CH_4.

The evolution of low molecular weight oligosiloxanes was not detected by MS, due to their high molecular weight, which is not measured by our instrument. However, Ionescu *et al.* observed with *in situ* FT-IR that these species correspond to octamethylsilsesquioxane, which evolutes in correspondence to the first two steps [Ionescu2009b]. Also the elimination of other cyclic and linear Si-O-Si monomers was observed with *in situ* FT-IR.

6.1.1.5.1. MK heat-treated at 500 °C

One of the important features of the silicon-based polymers and one of the reasons why they were chosen as promising materials for applications in LEDs is their thermal stability, which further increases after heat-treatment. The original idea for application in LEDs is the *in situ* molding, with the heat-treatment performed on the material positioned inside the chip. In order to crosslink the polymer *in situ*, the annealing temperature should be lower than the maximum temperature supported by the device, namely 300 °C. The lowest temperature at which the MK shows emission in the visible range, 500 °C, is far too high a temperature for direct processing. Nevertheless, *ex situ* shaping could also be an option. In this paragraph, the thermal stability of MK heat-treated at 500 °C is proven. The sample was subjected to thermal analysis, carried out both in air and argon, in order to define its thermal stability in different atmospheres (Figure 20).

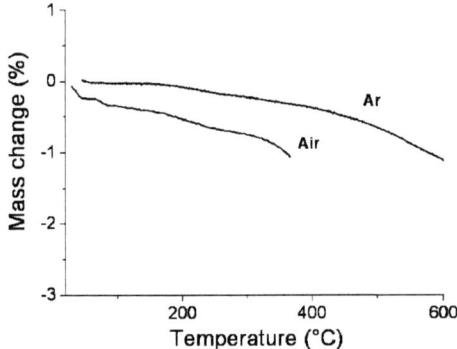

Figure 20: TGA analysis of the MK sample pre-annealed at 500 °C performed in air and argon.

The TGA of MK 500 °C was obtained in air only up to 366 °C. At this temperature 1% of the mass was lost. At the same temperature, the sample treated in argon displays only 0.3% mass loss, while 1% weight loss is reached for the sample in argon at 600 °C. The mentioned temperatures are far higher than the ones found in LED devices. Since an LED device could reach a maximum temperature of 150 °C during operation, the thermal stability of MK treated at 500 °C is largely satisfactory. Nevertheless, thermal analysis performed at the maximum reachable temperature of the device (150 °C) for a long time could provide additional information about the thermal stability of the sample.

6.1.1.6. XRD measurements

XRD analysis was performed on MK samples, in order to determine the possible formation of any crystalline phases during heating (Figure 21).

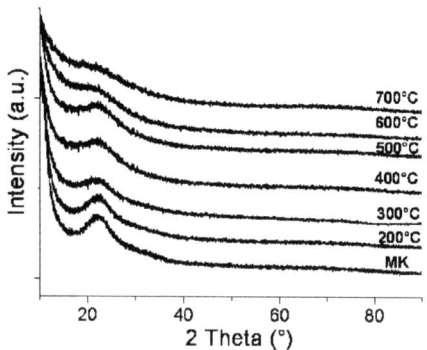

Figure 21: X-ray diffraction patterns of MK samples heat-treated at different temperatures.

Since no crystalline phase is detected with XRD measurements in any of the samples, we can conclude that MK polymer remains in the amorphous state after heat-treatment up to 700 °C.

6.1.1.7. Multinuclear Solid State MAS NMR spectroscopy (^{1}H, ^{13}C and ^{29}Si)

Multinuclear Solid State NMR is one of the most versatile methods for the characterization of insoluble polymers and pyrolysis intermediates. The method is very sensitive to the structural changes of the material.

In this paragraph, the thermal transformations of MK are analyzed by means of multinuclear MAS NMR. Liquid state NMR spectra of the MK polymer were published in the past [Schneider1988].

^{1}H, ^{13}C CP and ^{29}Si MAS NMR spectra of the MK polymer and MK heat-treated at 200, 300, 400, 500, 600 and 700 °C are shown in Figure 22 and their chemical shifts and bonds attribution are listed in Table 5.

6. Results and discussion

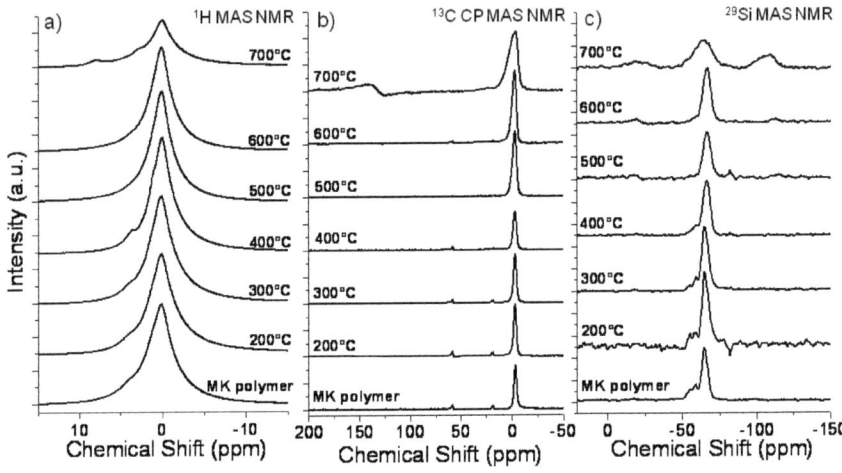

Figure 22: a) ^{1}H, b) ^{13}C CP and c) ^{29}Si MAS NMR spectra of MK heat-treated at different temperatures.

Table 5: Chemical shifts and relative bonds detected in MAS NMR spectra for MK samples heat-treated at different temperatures.

	^{1}H MAS NMR	^{13}C CP MAS NMR	^{29}Si MAS NMR
MK polymer	0.2 ppm (C-H); 3.7 ppm (–OCH_2–CH$_3$)	-3.6 ppm (CH_3); 18.5 ppm (H$_2$C-CH_3); 58.2 ppm (SiOC)	-64.8 ppm (SiO$_3$C); -58.4 ppm (SiO$_3$C); -54.9 ppm (SiO$_3$C)
MK 200 °C	0.2 ppm (C-H); 3.8 ppm (–OCH_2–CH$_3$)	-3.4 ppm (CH_3); 18.6 ppm (H$_2$C-CH_3); 58.3 ppm (SiOC)	-64.7 ppm (SiO$_3$C); -58.0 ppm (SiO$_3$C); -54.6 ppm (SiO$_3$C)
MK 300 °C	0.2 ppm (C-H); 3.8 ppm (–OCH_2–CH$_3$)	-3.5 ppm (CH_3); 18.5 ppm (H$_2$C-CH_3); 58.1 ppm (SiOC)	-64.9 ppm (SiO$_3$C); -58.6 ppm (SiO$_3$C); -54.4 ppm (SiO$_3$C)
MK 400 °C	0.1 ppm (C-H); 3.7 ppm (–OCH_2–CH$_3$)	-3.5 ppm (CH_3); 17.9 ppm (H$_2$C-CH_3); 58.0 ppm (SiOC)	-66.1 ppm (SiO$_3$C); -58.5 ppm (SiO$_3$C); -56.4 ppm (SiO$_3$C)
MK 500 °C	0.0 ppm (C-H); 3.0 ppm (–OCH_2–CH$_3$)	-2.0, -4.0 ppm (CH_3); 58.4 ppm (SiOC)	-66.1 ppm (SiO$_3$C); -19.3 ppm (SiO$_2$C$_2$); -110.1 ppm (SiO$_4$)

MK 600 °C	0.1 ppm (C-*H*); 3.6 ppm (–OC*H$_2$*–C*H$_3$*, unsaturated C*H$_2$*); 7.22 ppm (protonated carbon sp^2 (-*H*C$_{arom}$))	-2.1, -4.1 ppm (C*H$_3$*); 18.1 ppm (C*H$_2$*); 30.0 ppm (Si$_4$*C*); 58.0 ppm (*SiOC*)	-66.1 ppm (*Si*O$_3$C); -19.9 ppm (*Si*O$_2$C$_2$); -110.0 ppm (*Si*O$_4$)
MK 700 °C	0.0 ppm (C-*H*); 3.1 ppm (unsaturated C*H$_2$*); 7.8 ppm (protonated carbon sp^2 (-*H*C$_{arom}$))	-1.7, -2.9, -5.0 ppm (C*H$_3$*); 23.7 ppm (Si$_4$*C*); 136.0 ppm (*C*sp^2)	-63.9 ppm (*Si*O$_3$C); -19.3 ppm (*Si*O$_2$C$_2$); -107.1 ppm (*Si*O$_4$)

6.1.1.7.1. ^{1}H MAS NMR

^{1}H MAS NMR spectra of the MK polymer and MK heat-treated at 200, 300, 400, 500, 600 and 700 °C are shown in Fig 26 (a).

The MK polymer shows a peak at 0.2 ppm, denoting hydrogen bonded to carbon (C-*H*), when carbon is bonded to the silicon (Si-C*H$_3$*), and a small shoulder at around 3.7 ppm, corresponding to the hydrogen bonded to a carbon atom which is bonded to oxygen (ethoxy groups). These two peaks remain up to 500 °C. At 600 °C the main peak is at 0.1 ppm, the shoulder at 3.6 ppm grows in intensity, denoting at this temperature unsaturated CH$_2$, and another smaller shoulder arises at 7.2 ppm, which denotes the formation of protonated carbon sp^2 (-*H*C=). At 700 °C the main peak is still at 0.0 ppm; the peaks at 3.1 ppm and 7.8 ppm are relatively more intense.

6.1.1.7.2. ^{13}C CP MAS NMR

^{13}C CP MAS NMR spectra of MK samples are shown in Fig 26 (b). The untreated polymer and the samples heat-treated up to 400 °C show a main peak in the range -3.4 to -3.6 ppm, assigned to C*H$_3$*, and two small peaks in the range 17.9 to 18.6 ppm, related to H$_2$C-CH$_3$ bonds from ethoxy groups. The peaks at 58.0 to 58.3 ppm are related to [SiO*C*] units from Si-O-CH$_2$-CH$_3$. At 500 °C the main peak remains at -3.5 ppm and it is broadened, being simulated by two peaks (-2.0 and -4.0 ppm), indicating the gradually increasing number of Si-C containing species; the two small peaks at 17.9 to 18.6 ppm disappear and a very weak peak at 58.4 ppm can be observed, which corresponds to [SiOC] species. At 600 °C, the main peak is still at -3.3 ppm (-2.1 and -4.1 ppm in simulation); there are weak peaks at 18.1 ppm (CH_2), 30.0 ppm ([Si_4C]) and 58.0 ppm ([SiOC]). At 700 °C the main peak is slightly shifted (-4.3 ppm) and broadened, the band at 23.7 ppm ([Si_4C]) grows in intensity and another band at 136.0 ppm is detectable, denoting the formation of the carbon sp^2 (C=C).

6.1.1.7.3. ^{29}Si MAS NMR

^{29}Si MAS NMR spectra of MK polymer and MK heat-treated at 200, 300, 400, 500, 600 and 700 °C are shown in Fig 26 (c). The spectrum of the MK polymer shows the main peak at -64.8 ppm and two shoulders at -58.4 and -54.9 ppm, all indicating [SiO_3C] species, which denote the presence of ethoxy groups bonded to the silicon of the siloxane chain and also a ramified structure of the MK polymer (highly crosslinked). After treatment up to 400 °C, the NMR spectra do not show appreciable changes. At 500 °C, besides the main peak at -66.1 ppm, two new peaks develop at -19.3 and -110.1 ppm, related to [SiO_2C_2] and [SiO_4] species respectively, which were formed by means of crosslinking. At 600 °C the main peak related to [SiO_3C] is still present, and the peaks at -19.9 and -110.0 ppm increase in intensity. At 700 °C the main peak shifts to -63.9 ppm and broadens appreciably, and the two peaks at -19.3 and -107.1 ppm, present from 500 °C, significantly increase in intensity in relation to the main peak, indicating that the occurring of redistribution reactions.

6.1.1.7.4. Discussion of the Solid State MAS NMR results

In the case of MK polymer, the transition temperature from polymer to ceramic, as indicated by the formation of free carbon, was observed at 700 °C, as revealed in proton NMR (protonated carbon sp^2 at 7.9 ppm) and carbon CP NMR (136.0 ppm). The temperature increase from 500 to 700 °C results in enhanced degree of crosslinking of the MK polymer and differently coordinated Si atoms, namely [SiO_4], [SiO_3C] and [SiO_2C_2], as shown by the ^{29}Si NMR. The ramified structure is revealed by the presence and the broadening of their relative peaks, which denote almost all the coordination possibilities for SiOC materials. At 700 °C the broadening of the peaks is a further proof of the transition from polymer to ceramic.

6.1.1.8. EPR measurements

The fluorescence properties of the heat-treated MK were presented at the beginning of this section. The structural characterizations performed show how the polymeric structure undergoes a transformation into a ceramic network, and this occurs through bond cleavage and recombination processes associated with the processing temperature. Thermal analysis highlights mass losses during heat-treatment, which are accompanied by bond cleavages and formation of radicals. Thus, in the heat-treated polymers, the presence of dangling bonds (unpaired electrons, i.e. radicals or defects in the material) is expected. The following reactions, causing evolution of methane and hydrogen and formation of carbon and silicon radicals, were proposed by Bois *et al.* [Bois1994] and Radovanovic *et al.* [Radovanovic1999]:

$$\equiv Si-CH_3 \rightarrow \equiv Si\bullet + \bullet CH_3$$
$$\equiv C-H \rightarrow \equiv C\bullet + \bullet H$$
$$\bullet CH_3 + \equiv C-H \rightarrow \equiv C\bullet + CH_4$$
$$\bullet H + \equiv C-H \rightarrow \equiv C\bullet + H_2,$$

while the following by Trimmel *et al.* [Trimmel2003]:

$$\equiv Si-H \rightarrow \equiv Si\bullet + \bullet H$$
$$\equiv Si-CH_3 \rightarrow \equiv Si\bullet + \bullet CH_3$$
$$\equiv C-H \rightarrow \equiv C\bullet + \bullet H$$
$$\equiv Si-CH_3 + \bullet H \rightarrow \equiv Si\bullet + CH_4$$
$$\equiv C-H + \bullet CH_3 \rightarrow \equiv C\bullet + CH_4.$$

The cleavage of the bonds leads to the formation of carbon and silicon radicals and to methane and hydrogen evolution. Thus, silicon or carbon radicals are expected in the material.

As previously mentioned, in sol-gel materials, defects such as silicon and carbon dangling bonds, oxygen vacancies, non-bridging oxygen defects and radical carbonyl defects were found responsible for luminescence properties [Andronenko2006, Hayakawa2003, Baran2004, Prokes1998, Yang2001]. Correlation between luminescence intensity and radical concentration was previously observed in sol–

gel derived glasses and Si/SiO$_2$ materials. It was also established that the luminescence intensity increases linearly in dependence of the radical concentration [Hayakawa2003]. Thus, we hypothesize that the mechanism responsible for the luminescence of the MK samples is the presence of dangling bonds, as already observed in similar materials.

The identification of radicals in MK samples was carried out by means of X-band Electron Paramagnetic (Spin) Resonance (EPR/ESR) spectroscopy.

The measurements were first accomplished at room temperature. The EPR spectra obtained are characteristic for dangling bonds. An intense signal was detected for the sample heat-treated at 700 °C, but the signals relative to the remaining samples were very weak.

In order to increase the signal to noise ratio, the measurements were repeated at 15 K. Considering the Boltzmann distribution $n_{upper}/n_{lower} = \exp(-\Delta E/kT)$ (where n indicates the populations in the upper and lower energy states, ΔE the energy difference, k the Boltzmann constant and T the temperature), if the temperature is low, the upper energy state is less populated and the absorption of the microwave power results larger, as well as the EPR signal intensity. Therefore, at 15 K higher signal intensities were obtained for all the heat-treated MK samples (Figure 23, in grey). In order to determine g-factor, line width and spin concentration, the signal was numerically simulated using Matlab (Figure 23, in black).

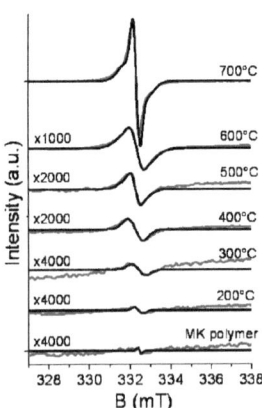

Figure 23: X-band EPR spectra (obtained at 15 K) of MK samples heat-treated at different temperatures (in grey) and their simulations (in black). The signals relative to the low temperature treated samples were magnified by a factor (labeled) for representative reasons.

6. Results and discussion

Also the low temperature EPR spectra of the MK samples show characteristic signals for dangling bonds. The quantitative analysis of the X-band EPR signals simulations provides g-factor values, line widths and spin concentrations of the radicals (Table 6).

Table 6: Quantitative analysis of X-band EPR signals simulations for MK samples.

Sample	g-factor	Line width (mT)	Concentration (spin·mg^{-1})
MK polymer	2.001	0.21	$2.68 \cdot 10^8$
MK 200 °C	2.001	0.5	$7.35 \cdot 10^8$
MK 300 °C	2.001	0.8	$1.59 \cdot 10^9$
MK 400 °C	2.002	0.87	$6.13 \cdot 10^9$
MK 500 °C	2.002	0.5 / 1.05	$8.49 \cdot 10^9$
MK 600 °C	2.002	0.65 / 1.38	$3.05 \cdot 10^{10}$
MK 700 °C	2.002	0.4 / 1.15	$1.88 \cdot 10^{13}$

The g-factor is usually composed by three elements (g_x, g_y and g_z), which are the diagonal elements of the g-tensor [Vij2006]. In our case, the g-tensor is isotropic ($g_x=g_y=g_z$), thus only one value is given in the table. In some samples, small anisotropy was present, however below the error. All observed species occur in a range of g-values of 2.001-2.002, therefore the paramagnetic spins are assigned to carbon species, in accordance with the results previously obtained for similar materials [Andronenko2006, Trassl2002, Berger2005]. Consequently, although the formation of silicon radicals during thermolysis was predicted, they are absent from the samples obtained after annealing of the preceramic polymer, as they probably form and react immediately to generate a bond. On the contrary, the carbon species seem to be more stable as radicals and can be detected in the samples by means of EPR analysis.

Three spectra, relative to MK heat-treated at 500-700 °C, consist of two overlapping resonances with identical g-value but different line widths. The identical g-factor indicates that both signals arise from carbon radicals. The line width denotes the hybridization state. The sp^2-hybridization is characterized by line widths larger than 1 mT and sp^3-hybridization by values below 1 mT, due to the different extents of spin delocalization within the "in-plane" dangling-bonds [Trassl2002, Prasad2000].

The line widths values relative to the EPR resonances of MK samples treated at different temperatures are displayed in Figure 24.

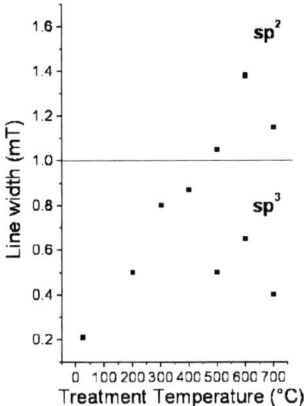

Figure 24: Line widths of the EPR resonances relative to the MK samples.

All heat-treated MK samples contain sp^3-type carbon dangling bonds, while sp^2-hybridized species are observed only in MK samples heat-treated from 500 °C. The sp^2-hybridized carbon, which appears at a temperature as low as 500 °C, is the precursor of C=C double bonds in a free carbon phase. None of the employed techniques could evidence the presence of these C=C bonds at 500 °C, but it cannot be directly discarded. Moreover, optical observations confirm the presence of a carbon phase because of the appearance of a yellow coloration at this temperature.

As mentioned in chapter 4, the area under the curve of the detected EPR signal is proportional to the number of radicals present. In order to compare the spin concentrations of the samples, this area must be normalized considering the weight of the material analyzed and the parameters used in the measurement. Furthermore, the spin concentration value was estimated by comparing the measured paramagnetic susceptibility with that of a spin standard of known spin concentration (DPPH).

The spin concentrations of the MK samples are illustrated in function of the treatment temperature in Figure 25 (the y-axis is in decimal logarithmic scale). Their exponential fit (in base 10) is also shown (linear fit in logarithmic scale).

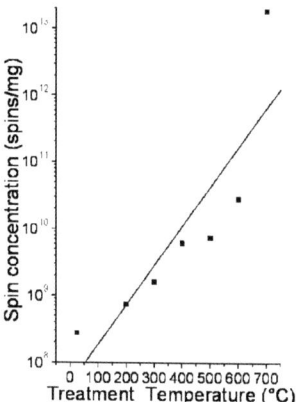

Figure 25: Concentration of dangling bonds in MK samples in function of the treatment temperature and fit of the exponential growth. Notice that the y-axis is in decimal logarithmic scale.

The radical concentration increases with the heat-treatment temperature. An exponential growth can be extrapolated, linear in logarithmic scale ((SD (standard deviation) = 0.87, R (correlation coefficient) = 0.87). The heat-treatment of the polysiloxane MK at 700 °C results in a spin concentration which is 3 to 4 orders of magnitude higher than those of the other samples. If the value corresponding to the MK annealed at 700 °C is excluded from the graph, a better linear fit in logarithmic scale is extrapolated (SD = 0.14, R = 0.99).

The spin concentrations of the MK samples plotted toward their fluorescence intensities, assumed as the integral of the fluorescence spectra in energy scale, are displayed in Figure 26. MK 700 °C is not shown for graphical reasons.

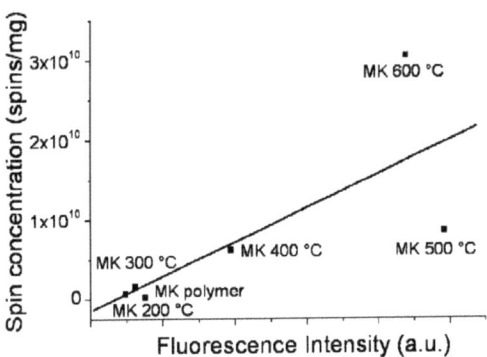

Figure 26: Correlation between spin concentration and fluorescence intensity of the MK samples.

Unlike the previously mentioned sol-gel luminescent materials, a linear correlation between spin concentration and fluorescence intensity cannot be detected for the MK samples. Dangling bonds are present in the materials and could contribute in some way to the luminescence emission; however they are definitely not the only mechanism. Coexistence of several mechanisms must be assumed for the fluorescence properties of the MK samples.

Although it is not possible to linearly correlate the spin concentration with the luminescence intensity, the EPR measurements supplied important information. Carbon radicals with sp^2-hybridization are detected at a temperature as low as 500 °C, denoting the starting point of the formation of sp^2 free carbon. The yellow color of the sample, in opposition to the white color of samples annealed at lower temperatures, further proves this statement. Other analysis methods (FT-IR, Raman and solid state NMR) were not able to determine this detail.

6.1.1.9. Discussion

In order to discuss the origin of the photoluminescence properties of the heat-treated MK samples, structural analyses as well as optical analyses must be considered.

The optical measurements performed on the MK samples heat-treated at different temperatures, i.e. the absorption spectra obtained by means of remission and reflection measurements and the fluorescence

6. Results and discussion

emission and excitation spectra, show a common trend, namely the red-shift of the absorption edges or maximum peaks with increasing annealing temperatures (Figures 5, 13 and 14). The simulations of the fluorescence spectra performed in Matlab indicate that new excitation and emission bands arise at increasing temperatures, in addition to the bands previously present at lower annealing temperatures. Therefore, new fluorescence centers appear as the annealing temperature increases. Since these centers are characterized by lower excitation and emission energies, the general effect on the spectra is the red-shift of the maximum peaks.

After treatment at temperatures up to 400 °C, the fluorescence emission spectra are in the UV range (Figure 5). After annealing up to 300 °C the spectra are almost identical and at 400 °C only slightly different. As structural analyses show, the MK structure slightly changes up to 400 °C. As attested for other silicon-based polymers, presented later on in the thesis, photoluminescence properties arise as the crosslinking proceeds. Therefore, a correspondence was found between crosslinking and photoluminescence properties: the crosslinking give probably rise to new chemical species with luminescence properties. The MK polymer is a highly crosslinked polysiloxane in the precursor state and is fluorescent. During annealing, it melts (50-60 °C) and further crosslinks. Since after annealing at 200 and 300 °C the fluorescence spectra are almost identical to those of MK polymer, there are no new luminescent species thermally created in comparison to those previously present in the precursor. The several emission peaks present in a single spectrum (for example in that of the MK polymer) could identify the presence of different emitters, and their intensity ratio changes as the thermal crosslinking proceeds. After annealing at 400 °C, the crosslinks thermally created become significant in relation to those present in the precursor and the fluorescence spectrum results slightly different from the previous ones: an increased contribution of the low energy bands was observed in both excitation and emission spectra, but no additional peaks were detected. This indicates that the luminescent species built through thermal crosslinking are previously existent in the polymer, thus their amount results increased and consequently the intensity of their emission peak. In summary, all MK samples are highly crosslinked and show fluorescence emission. A correlation between crosslinking and fluorescence emission is not excluded. However, if only MK samples are considered, this correlation is also not confirmed. In the next chapters, other silicon-based polymers will supply more proofs in support of this theory. Nevertheless, the photoluminescent species which develop with the crosslinking have not yet been clarified. Moreover, the luminescence analysis is complicated by the only partially clarified structure of

6. Results and discussion

the commercial polysiloxane MK and its development after annealing. Furthermore, impurities could be present in the commercial polymer.

Only after annealing at 500 °C, significant differences are achieved in the photoluminescence spectra (Figure 5). The samples heat-treated at 500, 600 and 700 °C show photoluminescence emission in the visible range. At 500 °C, sp^2 carbon radicals start to be detected from EPR analysis, indicating that the reactions leading to the formation of free carbon start at this temperature (Figure 24). The sp^2 carbon formation is concurrent with mineralization reactions involving methane and hydrogen evolution and considerable structural changes. At 500 and 600 °C, the sp^2 carbon is present in low amounts, detectable by EPR (in the form of its radical precursors) but not by solid state NMR analysis (Figure 22). The slightly yellow color of the samples after treatment at 500 and 600 °C attests the presence of sp^2 free carbon in low amounts. The brown color observed after treatment at 700 °C indicates the presence of sp^2 free carbon in considerable amounts, which is detected by solid state NMR analysis. Therefore, optical observations are important to confirm the sp^2 carbon presence and to appreciate its concentration.

The effect of the sp^2 free carbon is appreciable also in the absorption spectra (Figures 13 and 14). The absorption spectra relative to the samples annealed at 500 and 600 °C are wider than those obtained at lower temperatures and the absorption edge is shifted to the visible range. In the case of MK heat-treated at 700 °C, the absorption curve is appreciably more intense than those of the remaining samples, and the sample absorbs in the whole visible range up to 700 nm. As previously commented, absorption in the visible range implies coloration of the sample and the absorption measurements performed are in accordance with the colors of the samples. Considering that the sample coloration is dependent on the sp^2 carbon concentration, the free carbon concentration should be strictly correlated to the absorption of the samples. Therefore, we can attest that the position of the absorption edge of the samples heat-treated from 500 °C is exclusively dependent on the sp^2 free carbon concentration.

Although for absorption spectra a direct relationship was extrapolated between free carbon concentration and absorption edge, for the photoluminescence spectra the analysis is not so trivial. Heat-treatments from 500 °C imply emission spectra in the visible range. The new red-shifted peaks should be related to the formation of sp^2 carbon, but they could also be due to the structural changes occurring in the material. New bonds are formed during the evolution of methane and hydrogen, and the crosslinking proceeds further, creating new coordinations possibilities. Moreover, many dangling bonds are formed. The combination of these phenomena, all potentially responsible for the new red-shifted fluorescence emission peaks, results in relatively intense white emission. Nevertheless, since the red-shifted absorption edges of the polymers heat-treated from 500 °C are due to the sp^2 carbon formation and the excitation peaks of those samples are positioned in the same wavelength range of the red-shifted absorption peaks, it is clear that the visible luminescence must be due to the freshly formed sp^2 carbon structures. After heat-treatment at 600 °C, a similar emission as that obtained for the sample annealed at 500 °C is detected, only slightly red-shifted. After annealing at 700 °C, the emission spectrum is significantly further red-shifted; however the intensity decreases. This different behavior, occurring after temperature treatment only 100 °C higher, is imputable to the different concentration of sp^2 carbon developed. In low concentrations the sp^2 carbon contributes to the luminescence, while in high concentrations it quenches it (concentration quenching). This explains why in MK 700 °C the absorption, due to free carbon, is maximal, while the fluorescence is weaker.

The development of sp^2 carbon in the heat-treated polymers can be assumed to start in the form of double-bonded carbon which evolves in aromatic rings at higher temperature and in small agglomerations of aromatic rings at further high temperature. These structures start to form at 500 °C, as was proven by EPR measurements and confirmed by the yellow coloration of the sample and significantly develop at 700 °C. With further increase in temperature a carbon phase is formed.

Various similarities were observed between the photoluminescence properties of carbon nanotubes at different concentrations and the heat-treated MK samples, also characterized by sp^2 free carbon in different concentrations. In the case of the nanotubes, the photoluminescence intensity is dramatically reduced by aggregation of the isolated nanotubes [O'Connell2002]. Only separated nanotubes are fluorescent: their agglomeration causes the "inter-tube quenching" with decrease of luminescence emission. For this reason it is fundamental for nanotubes to be well dispersed [Wang2005]. The inter-tube quenching could be interpreted as concentration quenching, as it happens when the concentration of nanotubes is too high in a certain region of space. The nanotubes form bundles because of strong inter-tube van der Waals interactions, which perturb the electronic structure of the tubes and cause rapid excited-state relaxation [Wang2004]. Therefore, concentration quenching could be the reason of lack of visible luminescence in large sp^2 carbon structures, as for example graphite or coal.

In the case of the heat-treated MK, as the annealing temperature increases (500-600 °C), more and larger aromatic agglomerations with emissive properties form, until their concentration is high enough to create non-radiative pathways that quench the fluorescence properties (700 °C).

6. Results and discussion

Moreover, aggregation of nanotubes broadens the absorption spectra [O'Connell2002]. In Figure 27, the correlation between "polymer-bound" nanotubes concentration and absorbance spectra is reported from the literature (no aggregation is present) [Czerw2001].

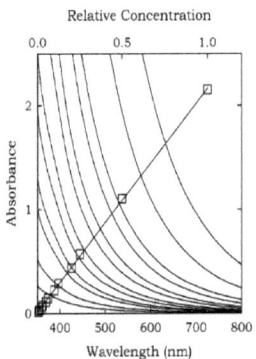

Figure 2. UV–vis absorption spectra as a function of concentration. The straight line shows a linear absorption dependence on concentration (scale on top).

Figure 27: Correlation between nanotubes concentration and absorbance spectra [Czerw2001].

Red-shifted absorption edges with increasing nanotubes concentration were found, similarly to the relationship previously extrapolated between free carbon concentration and absorption spectra in heat-treated MK samples. Large sp^2 carbon structures like graphite or coal can be considered at the extreme limit of this trend: the radiation is completely absorbed in the whole visible region and they are black.

Thus, in the case of MK, but also in other heat-treated silicon-based polymers reported later, the increase in annealing temperature generates new and larger aromatic agglomerations with photoluminescence properties, which at a certain temperature interact causing the quenching of the fluorescence emission.

After annealing at 500 and 600 °C, the conjugated carbon double bonds and the aromatic agglomerations absorb radiation and emit luminescence, like the nanotubes at low concentrations. At 600 °C both the absorption and emission spectra are slightly red-shifted compared to 500 °C, probably due to the presence of larger conjugation. At 700 °C the aromatic agglomerations grow further, and the concentration of sp^2-hybridized carbon is high enough to be detected in NMR measurements. The color of the sample is brown. The concentration of aromatic agglomerations is high enough to cause them to interact. The photoluminescence emission is further red-shifted, due to the larger carbon sp^2 structures. The absorption is stronger compared to lower annealing temperatures and in the whole 200-700 nm spectrum. Despite the more intense absorption, at this temperature the fluorescence intensity starts to decrease, due to the formation of non-radiative pathways, i.e. bridges between aromatic agglomerations, which quench the photoluminescence emission. For pyrolysis temperatures higher than 700 °C (not contemplated here), typically used for the PDC route, the MK samples become black (they strongly absorb in the whole visible range) and lose their luminescence properties. In the extreme case of large carbon structures, it was demonstrated that coal and coal derivatives show luminescence in the infrared [Friedel1966]. This could indicate a further red-shift of the emission.

Since analogies were observed between the heat-treated MK and carbon nanotubes in different concentrations, other fluorescence characteristics of the nanotubes will be mentioned, and compared with those of the MK samples.

The existence of extensive conjugated electronic structures, responsible for the luminescence of the nanotubes, was reported by Sun *et al.* and supports our findings [Sun2002]. Accordingly, the red-shift in the emission maximum of MK heat-treated at increasingly higher temperatures could be due to the increasing size of the conjugated structures, which could extend to aromatic agglomerations.

The luminescence of the nanotubes was also hypothesized to be due to the presence of defects in their structure which act as trapping sites for the excitation energy [Lin2005, Sun2002]. Accordingly, the dangling bonds observed in EPR measurements could be defects responsible for the photoluminescence properties of the sp^2 agglomerations in heat-treated MK. Nevertheless, in the light of our observations, this hypothesis is less probable for MK samples.

Furthermore, the nanotubes fluorescence is characterized by short lifetimes and it is due to the spin-allowed emission from singlet excitons [Wang2004]. Also the heat-treated MK samples showed quick emissions, although it was not possible to measure the lifetimes.

Finally, the fluorescence emission range of the nanotubes (500-600 nm) is similar to that of the aromatic agglomerations in heat-treated silicon-based polymers [Wang2007]. The different peaks in the visible emission spectra of MK samples annealed from 500 °C could indicate different sizes of the conjugated species, as different peaks in nanotubes fluorescence spectra indicate the different structures in the ensemble.

To conclude the discussion, it is plausible that the conjugated species developed in heat-treated MK have a role in luminescence emission. Not much is known about the structure of these freshly formed sp^2 structures. They could initially agglomerate in graphene sheets, which grow until they touch each other. In order to prove this hypothesis, a specific band for graphene sheets should be observed at 2670 cm^{-1} in Raman spectra, corresponding to the 2D band [Ni2008]. Unfortunately, in Raman spectra relative to MK samples no band relative to sp^2 carbon can be detected (Figure 16). Later on in the thesis, it will be shown that for other silicon-based polymers C=C bonds are visible in Raman spectra at relatively high treatment temperatures, but still no 2D band was detected. Thus, the formation of planar graphene sheets can be excluded. Aromatic agglomerations probably grow from the beginning in three directions, not only in two (graphene).

The findings are in agreement with the papers of Pivin et al., where the formation of carbon clusters provided the ceramic thin films with photoluminescence properties, which vanish with the growth of the carbon segregation [Pivin1998, Pivin2000]. However, in our case the carbon clusters are sp^2-hybridized not mainly sp^3. Moreover, affinities were found also with the previously mentioned papers of Green, Garcia and Lin et al., where the photoluminescence properties were related to the carbon impurities present in the glasses [Green1997, Garcia1995, Lin2000].

In conclusion, three main sources responsible for the fluorescence were discerned in the heat-treated MK samples: the crosslinking extent, the sp^2 free carbon and the carbon dangling bonds. Up to 400 °C, only crosslinking and radicals could contribute to the luminescence, since sp^2 carbon is not yet developed. An exponential growth in temperature was observed for the carbon radicals concentration. Nevertheless, no apparent changes are observed in photoluminescence spectra. Moreover, the graph displaying the correlation between luminescence intensity and spin concentration did not confirm a linear relationship. Therefore, although radicals have an effect on luminescence properties in similar materials, their role was not highlighted in MK samples heat-treated up to 400 °C. From 500 °C, sp^2 carbon, crosslinking and radicals could origin the fluorescence properties. From comparison with similar carbon materials, i.e. carbon nanotubes, and considering the absorption and fluorescence spectra, the free carbon was attested to be the main emitting source in the visible range. The role of crosslinking could not be discerned; nevertheless, as the correlation between crosslinking and luminescence was highlighted at low treatment temperatures and for different silicon-based polymers treated in this thesis, it should be one of the luminescence sources also for MK annealed from 500 °C. Again, the role of carbon radicals was not discerned also for these annealing temperatures. No linear relationship was found between luminescence intensity and spin concentration.

Consequently, the hypothesis of dangling bonds as main source of luminescence was abandoned. Although parallelisms and similarities between heat-treated MK and sol-gel materials were observed, the fluorescence mechanisms seem to be quite different, except for the formation of sp^2 carbon at higher annealing temperatures, phenomenon present in both systems [Green1997, Garcia1995].

In the following subchapters, different heat-treated silicon-based polymers will be presented. Although they are characterized by different structures, their luminescence behavior after heat-treatment can be brought back to that of annealed MK samples. Therefore, the present discussion is relevant also for the next photoluminescent systems.

Having partially explained the luminescence mechanisms, we consider the heat-treated MK polymer for application in LEDs. For heat-treatments up to 500 °C, the samples are not stable in air and reactions with moisture go on, causing an improvement in the luminescence intensity (Figure 11). This could be due to the further crosslinking of the samples. Application in protective atmosphere is an alternative solution. For heat-treatments from 600 °C, the luminescence does not vary after storage of the samples in air, indicating their stability.

Moreover, considering the heat-treatments in reducing atmosphere of the MK polymer, and the aromatic agglomerations as the luminescent species developing between 400 and 500 °C in argon atmosphere and between 500 and 600 °C in Ar/H$_2$ atmosphere, we can attest that the formation of free carbon is retarded if the heat-treatment is performed in reducing atmosphere (Figure 10). This is in accordance with the fact that the presence of hydrogen in the annealing atmosphere causes carbon loss from the sample [Liemersdorf2008]. The comparison between the heat-treatments of MK at 700 °C performed in the two different atmospheres corroborates this result.

6.1.2. Polydimethylsiloxane

In order to observe the role of crosslinking in the creation of luminescent species in MK polymer, two commercial polysiloxanes with defined structure were heat-treated and analyzed with luminescence measurements: the first polysiloxane is provided with non-crosslinkable substituents (polydimethylsiloxane), the second with crosslinkable substituents (polymethylvinylsiloxane). Both polymers are non-crosslinked polysiloxanes in the precursor state and their chemical structure is

6. Results and discussion

known. They were heat-treated and their luminescence was measured both in the precursor and heat-treated state. The two systems have not yet been fully analyzed and will be object of future studies.

The polydimethylsiloxane PDMS (SIGMA) does not show luminescence properties in the precursor state. Heat-treatment was applied at 500 °C, but no sample was recovered. Since this polymer is not able to crosslink, the linear chains evaporate. After heat-treatment at 300 °C, some of the polymer evaporated and some was still in the sample holder in liquid state (not crosslinked). The sample shows a weak photoluminescence (Figure 28).

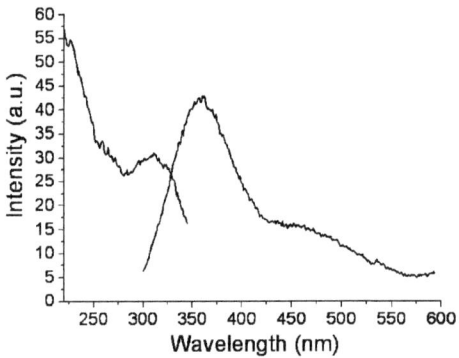

Figure 28: Fluorescence emission (bottom right) and excitation (top left) spectrum of the polydimethylsiloxane heat-treated at 300 °C, obtained with 250 nm excitation and 366 nm emission wavelengths, respectively.

Since no crosslinking takes place in polydimethylsiloxane, no emission is expected after heat-treatment, exactly as in the precursor. Nevertheless, a weak luminescence emission can be detected. The origin of this weak luminescence is not clear, as it does not derive from luminescent species created by the crosslinking. A possibility could be the formation of dangling bonds created during annealing.

Anyway, this emission is irrelevant compared to the strong emission shown by MK samples. Therefore, these results do not contradict the hypothesis of crosslinking as cause of photoluminescent species in silicon-based polymers. Moreover, if the only source of luminescence in PDMS is the presence of dangling bonds, their role was demonstrated to be marginal, as previously proven in the MK samples.

The next step is the observation of the luminescence behavior of a polysiloxane that is able to crosslink (polymethylvinylsiloxane), in order to correlate the photoluminescence properties with the crosslinking.

6.1.3. Polymethylvinylsiloxane

Two commercially available polymethylvinylsiloxanes, namely polyvinylmethylsiloxane PVMS homopolymer (VMS-005, cyclic, ABCR) and vinylmethylsiloxane VMS homopolymer, linear (VMS-T11, linear, ABCR), were annealed at different temperatures up to 600 °C and their photoluminescence properties were analyzed. Their molecular structure is shown in Figure 29:

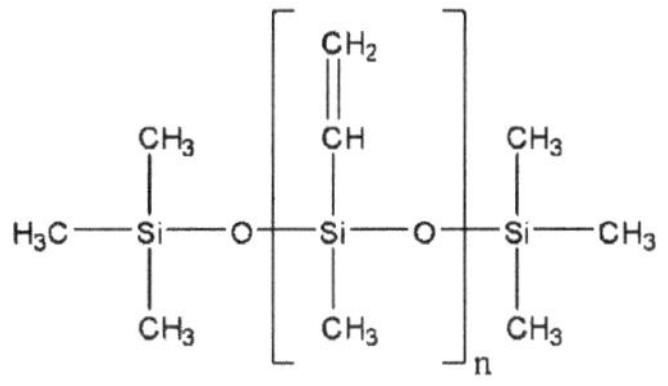

Figure 29: Molecular structural unit of the PVMS and VMS polymers. The PVMS is cyclic, the VMS is linear.

In Figure 30, the maximum excitation and emission spectra relative to PVMS (a) and VMS (b) samples heat-treated at different temperatures, obtained with emission and excitation wavelengths according to Table 7, are shown.

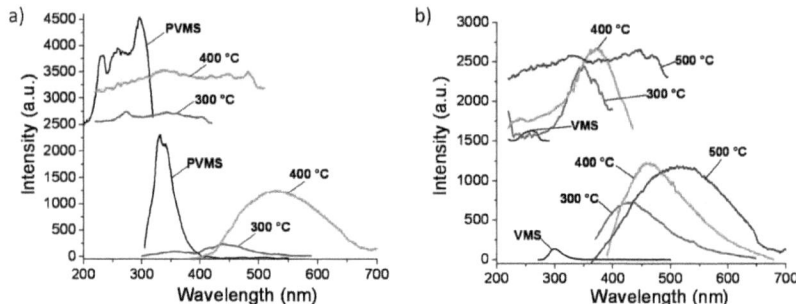

Figure 30: Fluorescence emission (bottom right) and excitation (top left) spectra of the cyclic (PVMS) (a) and linear (VMS) (b) polymethylvinylsiloxanes, heat-treated at different temperatures.

Table 7: Excitation and emission wavelengths used to obtain fluorescence spectra of the PVMS and VMS samples. They correspond to the maximum emission and excitation peaks.

Sample	Excitation wavelength (nm)	Emission wavelength (nm)
PVMS	297	328
PVMS 300 °C	274	438
PVMS 400 °C	333	531
VMS	259	298
VMS 300 °C	325	431
VMS 400 °C	370	457
VMS 500 °C	334	514

After heat-treatment at 200 °C, no PVMS sample was recovered, while for VMS liquid sample was recovered, but the fluorescence intensity was too low to be illustrated in the graph. Moreover, PVMS heat-treated at 500 °C and VMS heat-treated at 600 °C did not show fluorescence emission, due to the free carbon quenching.

In both polymers, vinyl groups provide crosslinking possibilities and the samples show photoluminescence properties. The emission ranges are red-shifted as the annealing temperature increases, as previously observed for MK samples.

Unexpectedly, also the starting polymers show fluorescence emission. The origin of this emission is not clear, because no luminescence centers are present in the precursors. The only difference between PVMS and PDMS is the presence of vinyl groups. However, fluorescence emission from vinyl groups was not reported so far. Impurities in the commercial polymers could also originate photoluminescence properties. Annealing at 200 °C is not effective in order to crosslink the polymers. PVMS evaporates completely, while VMS remains liquid. The reason of the quenching of the VMS fluorescence emission after annealing at 200 °C is not clear. For annealing starting from 300 °C, crosslinking takes place in both polymers and red-shifted emissions in respect to the starting polymers are detected. As in MK, at higher heat-treatment temperatures the formation of free carbon has an effect on the luminescence properties. Structural analyses have not been yet performed on these polymers and will be investigated in the future.

In conclusion, the effect of the crosslinking on the fluorescence properties was demonstrated. Moreover, a new source of luminescence was identified, not related to the crosslinking, but its origin is not clear. The clarification of the relationship between photoluminescence and crosslinking is still in an early stage. In the next chapters, the role of crosslinking on fluorescence properties will be outlined in other silicon-based polymers.

6.2. Polysilazanes

6.2.1. KiON Ceraset® PUVMS

The polyureamethylvinylsilazane (KiON Ceraset® PUVMS) was initially chosen because of its commercial availability, low cost, mouldability, crosslinkability and non-toxicity [KiON]. Its stability against oxygen makes it suitable for handling in air.
This patented thermosetting resin contains repeated units, in which silicon and nitrogen atoms are bonded in an alternating sequence, having both cyclic and linear features. Upon contact with air's moisture slight hydrolysis of the polymer can occur, with generation of ammonia [KiON]. A small percentage of urea functionality is present in the polymer. Urea- or isocyanate-containing compounds were reported to improve the chemical and physical properties of the polysilazanes [Li2001].
The KiON Ceraset® PUVMS is a low viscosity liquid polysilazane. It crosslinks by heating at 180-200 °C or at lower temperatures if a small quantity of a free radical initiator is added (peroxide) or by exposure to UV radiation in the presence of a UV sensitizer [KiON]. Upon curing, Ceraset is converted into a rigid solid, insoluble in common organic solvents, water and dilute acids and bases. Crosslinking of Ceraset at 280 °C leads to optically transparent and completely crosslinked bulk materials [Li2001, Li2000a].
Pyrolysis under inert atmosphere of Ceraset leads to amorphous (up to 1400 °C) or crystalline (above 1400 °C) SiCN ceramics [KiON, Janakiraman2009]. Depending on the pyrolysis atmosphere, silicon carbide (argon, nitrogen), silicon nitride (nitrogen, ammonia), SiCN (argon) and SiCNO (air) systems can be created [KiON]. KiON Ceraset Polyureasilazane was designed for Ceramic Matrix Composites (CMCs), Metal Matrix Composites (MMCs), corrosion and oxidation resistant and high temperature coatings, infiltrants for ceramic performs and other high performance ceramic precursor applications [KiON]. Several publications can be found in literature about Ceraset, Ceraset SN and Ceraset Ultra, as suitable precursors for PDCs [An2004, Bahadur2003, Bakumov2007, Bitterlich2005, Brahmandam2007, Cross2006a, Cross2006b, Das2004, Dhamne2005, Duan2004, Duan2005, Gao2008a, Gao2008b, Garcia2003, Gasch2001, Gudapati2006, Nagaiah2006, Hanemann2002,

6. Results and discussion

Hauser2008, Janakiraman2006, Klaffke2006, Kamperman2004, Kamperman2007, Kamperman2008, Katsuda2006, Kojima2002, Li2007, Li2008, Liew2001, Liu2002, Mackin2000, Markgraaff1996, Plovnick2000, Rak2001, Reddy2003, Ryu2007, Saha2003a, Saha2003b, Schulz2004, Seok1998, Shah2002, Shah2005, Stantschev2005, Wan2001, Wan2002, Wan2005, Wan2006, Wang2008, Yang2004a, Yang2004b, Yang2005, Yang2006, Yang2007, Yang2008a, Yang2008b, Yang2008c, Youngblood2002, Zemanova2002, Vakifahmetoglu2009, Pham2007].

Recently, ceramic SiCNO systems were found to be luminescent [Morcos2008a, Ferraioli2008].

As in the case of MK polymer, also the structure of Ceraset was not completely clarified. The structure provided by the KiON Corp. (Clariant) and reported in many publications is illustrated in Figure 31.

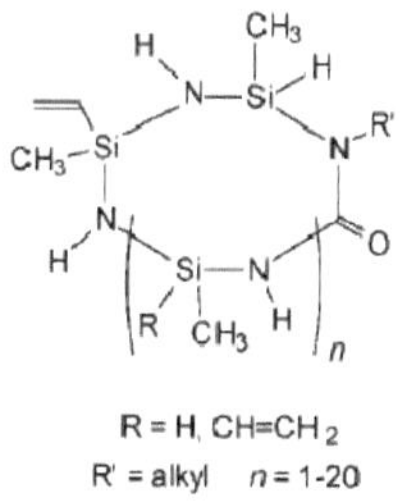

$R = H, CH=CH_2$
$R' = alkyl \quad n = 1-20$

Figure 31: Molecular structure of KiON Ceraset® PUVMS as provided by KiON Corp.

Some organic substituents have not been yet clarified. Therefore, we further investigated the structure in the present study [Menapace2008]. The polymer possesses crosslinkable substituents, as vinyl and Si-H groups. Also the N-H groups contribute to the crosslinking, through dehydrocoupling and transamination reactions. Ceraset is composed by a mixture of low molecular weight oligomers and high molecular weight polymeric chains.

The liquid Ceraset was heat-treated at 200, 300, 400, 500 and 600 °C for 2 hours under Ar flow, with heating rate 50 °C/h and free cooling. Solid crosslinked samples were obtained, which retained the shape of the mold, and were subsequently powdered.

6.2.1.1. Photoluminescence measurements

6.2.1.1.1. Fluorescence measurements

Photoluminescence properties were observed for all heat-treated samples [Menapace2008]. Figure 32 displays the excitation and emission spectra of Ceraset annealed at different temperatures. The excitation and emission wavelengths used for the measurements correspond to the maximum intensity peaks (Table 8).

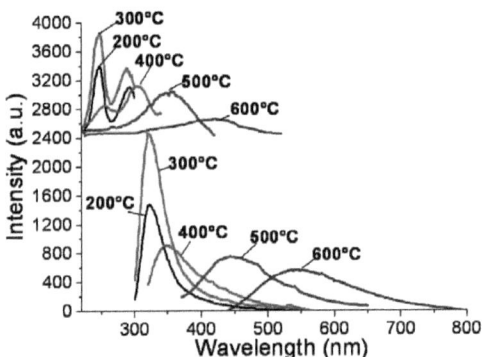

Figure 32: Fluorescence emission (bottom right) and excitation (top left) spectra of the Ceraset samples heat-treated at different temperatures (labeled). The excitation and emission wavelengths applied are described in Table 8.

Table 8: Excitation and emission wavelengths used to obtain the fluorescence spectra for polyureamethylvinylsilazane Ceraset samples. They correspond to the maximum emission and excitation peaks.

Sample	Excitation wavelength (nm)	Emission wavelength (nm)
Ceraset 200 °C	247	319
Ceraset 300 °C	247	319
Ceraset 400 °C	304	350
Ceraset 500 °C	350	442
Ceraset 600 °C	424	543

Also the liquid precursor shows weak luminescence. For representative reasons, the fluorescence emissions of the untreated Ceraset obtained at 250 and 360 nm as the excitation wavelengths are shown in a separate graph (Figure 33).

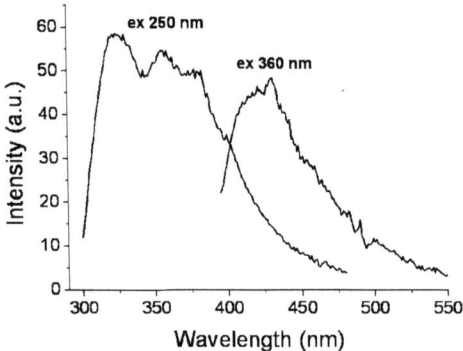

Figure 33: Fluorescence emissions of liquid Ceraset obtained with excitations of 250 and 360 nm.

The emission intensity of the liquid Ceraset is very low, though luminescence properties can be detected.

The samples crosslinked at 200 and 300 °C are transparent and colorless and show luminescence emission in the UV range. Their emission and excitation peaks are identical in shape, but Ceraset 300 °C shows increased intensity compared to Ceraset 200 °C. Excitation peaks were detected in the range between 234 nm and 326 nm. The main emission range is detected between 310 nm and 379 nm with maximum at 319 nm.

The sample heat-treated at 400 °C shows slightly red-shifted excitation and emission ranges compared to the previous samples, with maximum emission at 350 nm. The sample shows the same excitation peaks as the previous samples and an additional emission peak at 350 nm. The emission intensity is significantly decreased in comparison to samples annealed at lower temperatures.

The sample heat-treated at 500 °C presents excitation peaks between 299 nm and 366 nm, while emission peaks appear between 386 nm and 563 nm with maximum at 431 nm.

The annealing of Ceraset at 600 °C results in a sample exhibiting lower luminescence intensity in comparison to the previous sample (calculated as the integral under the spectrum in energy scale), visible also from the q.e. results reported later and the optical observations. The intensity decrease at 600 °C is due to the development of free carbon, as attested by MAS NMR spectra (see later), similarly to MK annealed at 700 °C. At this temperature, an additional excitation peak at 416 nm can be identified in respect to the previous sample and the emission range appears between 527 and 620 nm.

As clearly visible in Figure 32, by increasing the temperature from 300 to 600 °C, red-shifted excitation and emission ranges are detected.

Powdered Ceraset samples annealed at different temperatures are shown in Figure 34, under white (a) and UV (360-400 nm) (b) light.

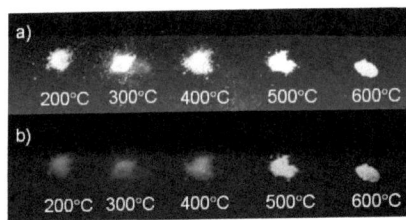

Figure 34: Photographs representing powdered Ceraset samples annealed at different temperatures under white (a) and UV (360-400 nm) (b) light, respectively. For colored pictures the reader can refer to http://tuprints.ulb.tu-darmstadt.de/2085/.

In Figure 35, the emission spectra of the Ceraset samples obtained with excitation wavelength of 250 nm (a) and 360 nm (typically used for LED applications) (b), are shown.

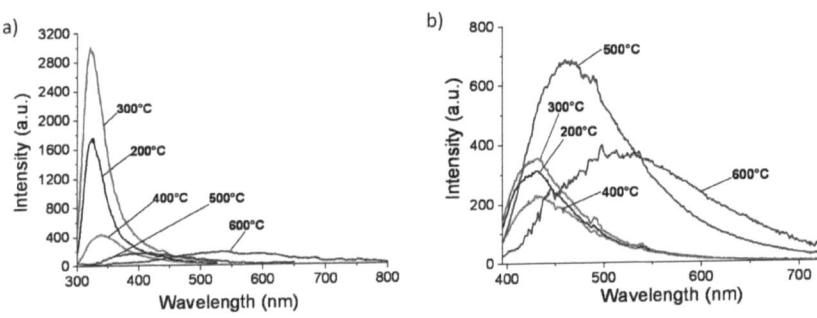

Figure 35: Fluorescence emission spectra of the Ceraset samples heat-treated at different temperatures obtained with excitation wavelength of 250 nm (a) and 360 nm (b).

With excitation of 250 nm, the Ceraset heat-treated at 300 °C shows the most intense fluorescence emission. The emission intensity decreases for treatments at higher temperatures. This is in accordance with the excitation spectra of Figure 32, where the excitation peak at 247 nm is maximal at 200 and 300 °C, decreases at 400 °C and disappears at higher temperatures. On the contrary, 360 nm corresponds to the maximum excitation peak of Ceraset 500 °C: therefore this sample shows the most intense emission spectrum with this excitation wavelength. Ceraset heat-treated at 500 °C is an interesting material for LED applications, because the whole emission spectrum is in the visible range and the excitation maximum is at around 360 nm. The material is transparent and the processing temperature is significantly decreased compared to the molding temperature of rare earth doped glasses [Zhu2007, Sun2007].

In Figure 36, Ceraset heat-treated at 500 °C (bulk sample) is shown under white (a) and UV (360-400 nm) (b) light.

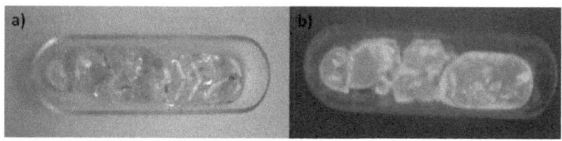

Figure 36: Photographs representing Ceraset annealed at 500 °C under white (a) and UV (360-400 nm) (b) light, respectively. For colored pictures the reader can refer to http://tuprints.ulb.tu-darmstadt.de/2085/.

The sample is transparent in both cases, slightly yellow under white light and emitting blue-greenish light under irradiation of UV light.

Figure 37 represents Ceraset annealed at 600 °C, under white (a) and UV (360-400 nm) (b) light, respectively.

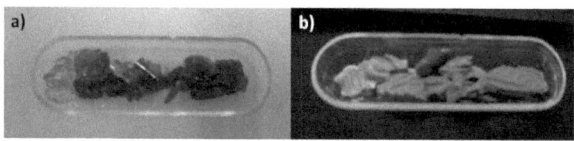

Figure 37: Photographs representing Ceraset annealed at 600 °C under white (a) and UV (360-400 nm) (b) light, respectively. For colored pictures the reader can refer to http://tuprints.ulb.tu-darmstadt.de/2085/.

6. Results and discussion

The sample is orange-red and translucent under white light, and completely opaque under UV light.
As in the case of MK polymer, the annealed Ceraset samples are transparent and colorless up to 400 °C. At 500 °C they acquire a yellow coloration and show visible luminescence if irradiated with UV light, remaining transparent in both cases; at 600 °C they are translucent red under white light and turn opaque under UV light.
The spectral emissions obtained with 360 nm as the excitation wavelength of the Ceraset samples heat-treated at 200-600 °C are represented on the CIE chromaticity diagram 1931 (Figure 38).

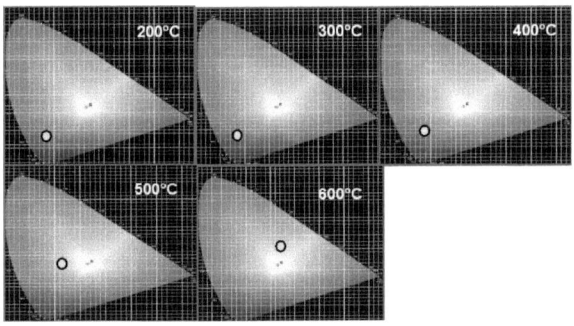

Figure 38: CIE chromaticity diagrams 1931 of the Ceraset samples annealed at 200-600 °C. For colored pictures the reader can refer to http://tuprints.ulb.tu-darmstadt.de/2085/.

The color coordinates are in accordance with the colors of the samples under UV light (Figure 34, 36 and 37).

6.2.1.1.2. Stability of the fluorescence during storage

In order to monitor the change in fluorescence properties of the Ceraset samples following storage in air, the fluorescence spectra were detected within the first week after the heat-treatment and repeated two years later, both obtained with 250 nm as the excitation wavelength (Figure 39).

6. Results and discussion

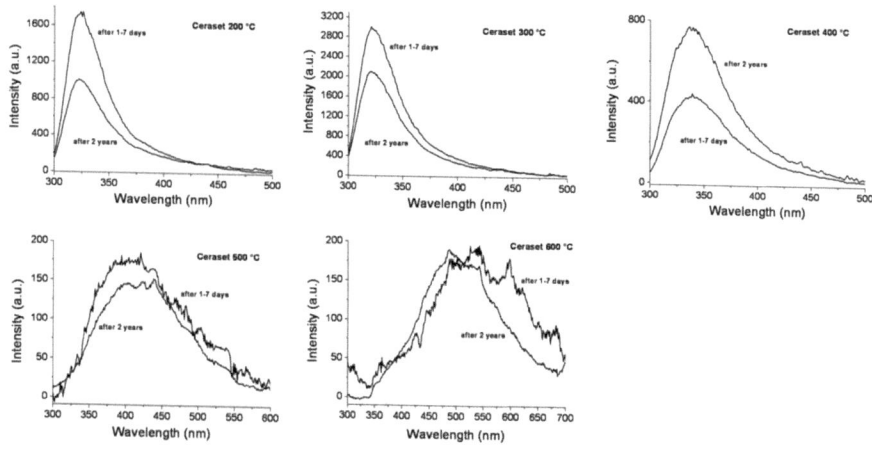

Figure 39: Fluorescence spectra of the Ceraset samples obtained during the first week after heat-treatment and 2 years later (at 250 nm excitation).

The fluorescence intensities of Ceraset samples treated at 200 and 300 °C decrease after two years of storage in air, that of sample Ceraset 400 °C increases, while for Ceraset 500 and 600 °C similar intensities are detected, with small differences. The spectra do not evidence the rise of new peaks and mainly maintain the same shape in time. In Ceraset 500 and 600 °C, the ratio scattering/fluorescence is quite high and the differences between the spectra could be due to errors coming from the subtraction of the scattering. The exposure to air and moisture of polysilazanes can cause hydrolysis reactions with loss of ammonia.

6.2.1.1.3. Quantum efficiency

The quantum efficiency of Ceraset heat-treated at 500 °C was measured with an integrating sphere and resulted to be 7%. The quantum efficiencies of the remaining Ceraset samples were estimated by comparing the integral of their emission spectra (in energy scale) with the integral of the emission spectra of samples of known quantum efficiency, with emission in the same range (Table 9).

Table 9: Quantum efficiencies of the Ceraset samples.

Sample	Quantum efficiency (%)
Ceraset 200 °C	16.6*
Ceraset 300 °C	28.6*
Ceraset 400 °C	14.9*
Ceraset 500 °C	7.0
Ceraset 600 °C	3.0

The samples labeled with * emit in the UV range (<394 nm), for which no measurement was successful; therefore the values were calculated on comparison with a sample with emission maximum at 394 nm.

6.2.1.2. Absorption measurements

Additional information about the electronic transitions occurring in the Ceraset samples and their variation after temperature treatment can be obtained with absorption measurements.

6.2.1.2.1. UV-Vis-NIR spectroscopy

As in the case of MK, standard UV-Vis measurements in THF solvent were performed on the Ceraset samples. However, the solubility decreases as the treatment temperature increases. Only liquid Ceraset was soluble and its absorption spectrum was measured in THF solution (Figure 40). UV-Vis measurements were previously carried out on Ceraset [Li2000a].

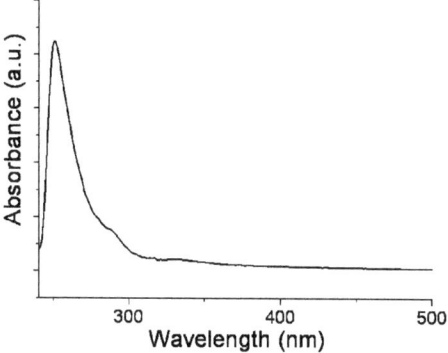

Figure 40: UV-Vis absorption spectrum of liquid Ceraset.

The absorption edge of Ceraset occurs at around 300 nm (4.13 eV). No comparison with crosslinked samples was possible.

6.2.1.2.2. Remission measurements

The absorption/scattering curves of the Ceraset samples obtained with remission measurements are presented in Figure 41.

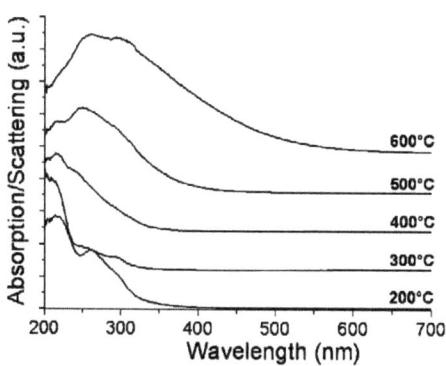

Figure 41: Absorption/scattering spectra of Ceraset samples obtained with remission measurements.

The absorption spectra obtained by means of remission measurements show red-shifted absorption edges as the heat-treatment temperature increases, in accordance with fluorescence emission and excitation spectra and with the absorption spectra obtained for MK samples. Moreover, as the annealing temperature increases, increasingly intense absorption spectra are obtained. As previously explained in the discussion on the MK polymer, the absorption of the samples annealed starting from 500 °C is due to the development of free carbon, which shifts the absorption edge to the visible range. Comparably to MK heat-treated at 700 °C, the absorption curve of Ceraset 600 °C is significantly more intense than those of the other samples, while the fluorescence emission starts to decrease at this temperature. This is due to the development of free carbon in large amounts, which causes improved absorption but quenching of the emission due to its high concentration. The sample annealed at 200 °C is an exception, probably related to its partially liquid structure.

6.2.1.2.3. Reflection measurements

The reflection spectra were obtained using synchronous excitation and detection wavelengths with the Cary Eclipse Varian spectrometer. The absorption/scattering spectra obtained from the reflection data relative to the Ceraset samples (using the Kubelka-Munk function) are presented in Figure 42.

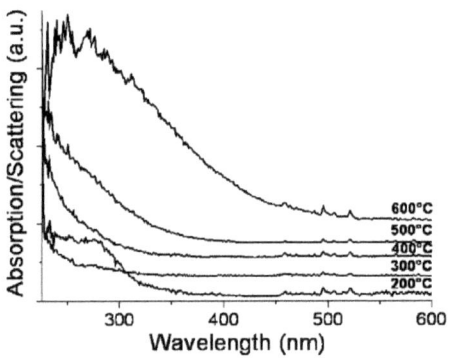

Figure 42: Absorption/scattering spectra relative to the Ceraset samples obtained with reflection measurements.

Except the sample heat-treated at 200 °C, the absorption edge of the Ceraset samples shifts toward less energetic wavelengths as the treatment temperature increases.

The samples annealed at 300 and 400 °C absorb radiation up to about 330 nm (3.75 eV), after heat-treatment at 500 °C the absorption edge is at 400 nm (3.10 eV), while for annealing at 600 °C at 550 nm (2.25 eV). The absorption spectra are in agreement with the excitation spectra of Figure 32, with the remission spectra and with optical observations. Moreover, the results show the same trend as the photoluminescence emissions. The sample annealed at 200 °C is not completely reflective and transmits some radiation in the whole spectral range. Moreover, it shows an unusual behavior, which could be attributed to its partially liquid structure. The reflection measurements are in accordance with the KiON description of the product, where the UV absorption edge of the cured Ceraset resin (at about 280 °C) occurs at around 300 nm [KiON].

6.2.1.3. FT-IR spectroscopy

The FT-IR analysis of Ceraset and heat-treated Ceraset was presented in the literature [Li2001, Li2000a, Dhamne2005, Kojima2002]. The analysis of Ceraset SN [Shah2002] and Ceraset annealed in nitrogen and ammonia atmospheres [Wan2002] are also reported. Nevertheless, not all the treatment temperatures here studied are published and a smaller heating rate was used in this work for the annealing. In Figure 43, the FT-IR spectra relative to the Ceraset samples are presented. In Table 10, the vibration bands and respective bonds are listed. The liquid Ceraset was measured using the ATR device, while the heat-treated samples were measured using KBr discs.

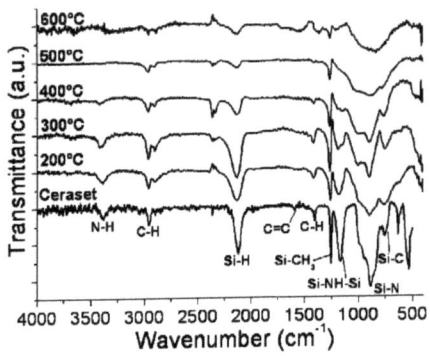

Figure 43: FT-IR spectra of the Ceraset samples.

Table 10: Vibration bands and respective bonds detected by means of FT-IR analysis of the Ceraset samples.

Bond assignment	Vibration band (cm^{-1})
$v_s(N\text{-}H)$	3380
$v_{as}(\text{-}CH_3)$	2960
$v_{as}(CH)$	2902
$v(Si\text{-}H)$	2119
$v(C=C)$	1592
$\delta_{as}(C\text{-}H)$	1412
$\delta_s(Si\text{-}CH_3)$	1254
$v(Si\text{-}NH\text{-}Si)$	1166
$v_{as}(Si\text{-}N)$	889
$\delta_s(Si\text{-}C)$	758
$v(C=C)$	633

After heat-treatment at 200 °C, the C=C bonds from the vinyl substituents, slightly visible in the precursor, have disappeared; Si-H and N-H bonds are unreacted. Also at 300 °C, Si-H, N-H and C-H peaks are not reduced, but the band relative to Si-NH-Si starts to decrease in intensity. At 400 °C, the Si-H and N-H peaks are strongly reduced and also the C-H band starts to decrease. At 500 °C, the peaks corresponding to Si-C and Si-N broaden significantly; N-H groups are not present anymore. At 600 °C, the peaks relative to Si-N and Si-C are further broadened.

At a temperature as low as 300 °C the crosslinking reactions through hydrosilylation and polymerization of vinyl groups are proven by means of FT-IR analysis, highlighted by the disappearance of vinyl groups. At 400 °C the reduced number of N-H and Si-H groups indicates the occurrence of transamination reactions and dehydrocoupling. At temperatures starting from 500 °C mineralization reactions take place. The Si-CH$_3$ and Si-NH-Si peaks significantly decrease in intensity as the treatment temperature increases, indicating the mineralization of the polymer and the occurred crosslinking via dehydrocoupling and transamination reactions, respectively. The findings are in accordance with the literature [Li2001]. Nevertheless, in our analysis it was not possible to detect bands related to C-H vibrations from vinylsilyl groups (H$_2$C=CH-Si) at 3048 and 3007 cm^{-1}, because of the method limitations in our measurements.

6.2.1.4. Raman spectroscopy

The heat-treated Ceraset samples were investigated with confocal micro-Raman spectrometer (Figure 44 and Table 11). Raman spectra of Ceraset crosslinked at different temperatures are present in the literature, but not for all the annealing temperatures analyzed here and not for the same heating rate of the annealing [Li2001]. In our measurements, an excitation laser of 633 nm was found optimal in order to minimize the fluorescence background.

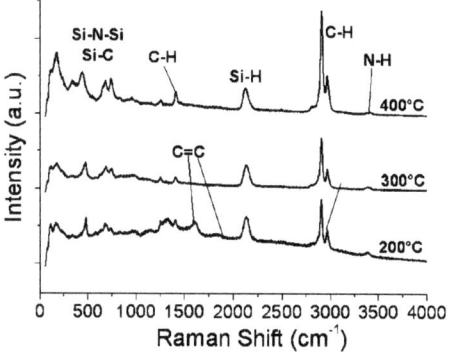

Figure 44: Raman spectra of the Ceraset samples using 633 nm as laser wavelength.

Table 11: Vibration bands and respective bonds detected by means of Raman analysis of the Ceraset samples.

Bond assignment	Raman shift (cm^{-1})
$v(N-H)$	3386
$v_{as}(-CH_3)$	2960
$v_{as}(CH_2)$	2899
$v(Si-H)$	2129
$v(C=C)$	1818
$v(C=C)$	1588
$\delta(C-H)$	1403
$v_s(Si-C)+v(Si-N-Si)$	164-944
$v(Si-N)$	475

The C=C bonds, visible in the sample heat-treated at 200 °C, disappear at higher temperatures. They correspond to the vinyl groups, which are still present at 200 °C (the sample is only partially crosslinked), but completely react via hydrosilylation or vinyl polymerization at higher temperatures. At increasing treatment temperature, except for the disappearance of C=C bonds, no variations in the bands can be detected. The N-H, the Si-H and the C-H groups as well as the Si-N-Si and Si-C bonds are present up to 400 °C. For heat-treatments from 500 °C, the fluorescence of the samples interferes with the Raman measurements and the bands are partially (500 °C) or almost totally (600 °C) hidden (not shown).

The samples heat-treated at higher temperatures (400-600 °C) were also analyzed with the IR/Raman spectrometer Bruker IFS 55 - FRA 106, with laser wavelength 1064 nm. The spectra are reported in Figure 45.

6. Results and discussion

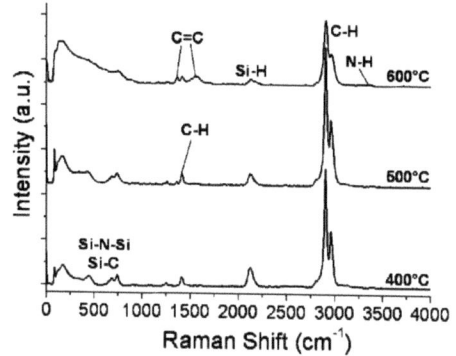

Figure 45: Raman spectra of the Ceraset samples heat-treated at 400-600 °C using 1064 nm as laser wavelength.

In this case, it is possible to avoid the luminescence interference for samples annealed up to 600 °C. The N-H, the Si-H and the C-H groups as well as the Si-N-Si and Si-C bonds are now clearly detectable up to 600 °C. The Si-H peak decreases in intensity as the treatment temperature increases, due to the occurrence of hydrosilylation and dehydrocoupling reactions. Also N-H groups are reduced compared to lower temperatures due to dehydrocoupling and transamination reactions. Two new bands with maxima at 1360 and 1555 cm^{-1} appear at 500 °C and increase in intensity at 600 °C, which are more clearly represented in Figure 46.

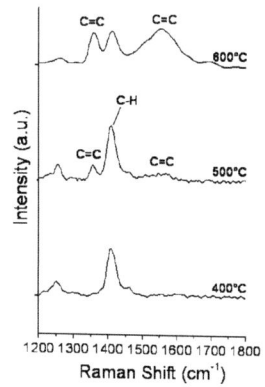

Figure 46: Raman spectra in the range 1200-1800 cm^{-1} of the Ceraset samples heat-treated at 400-600 °C using 1064 nm as laser wavelength.

The two new bands appear in the ranges of the Raman shifts relative to the D and G bands, which are detected in polymer derived ceramics after pyrolysis at higher temperatures. Nevertheless, with laser excitation at 1064 nm, the D band is reported to appear at around 1290 cm^{-1} [Larsen2006, Dresselhaus2000]. Therefore, the two new peaks cannot be assigned to the D and G bands. However, they denote C=C double bonds, which indicate the formation of sp^2 free carbon, in accordance with optical observations. The peak positions are well-matched with aromatic ring chain vibrations (1580, 1600 cm^{-1}) and in particular with the Raman shifts relative to polycyclic aromatic hydrocarbons (naphthalene, anthracene, pyrene and perylene) [Shinohara1998]. The broadness of the band with maximum at 1555 cm^{-1} could indicate the presence of aromatic agglomerations of different sizes. In addition, C=N bonds are also detectable in the range 1610-1680 cm^{-1}. In the Raman spectra of Ceraset published by Li *et al.* the sp^2 carbon is not detectable up to 800 °C, probably because of the higher heating rates used in the pyrolysis [Li2001].

6.2.1.5. TGA/DTG/MS

Thermal analysis was carried out on Ceraset in Ar atmosphere and the TGA- and its differential DTG-curves are represented in Figure 47. TGA curves of Ceraset were previously published, obtained in argon [Li2001, Janakiraman2006, Li2008, Seok1998] and nitrogen [Li2007, Shah2002 (Ceraset^{SN})].

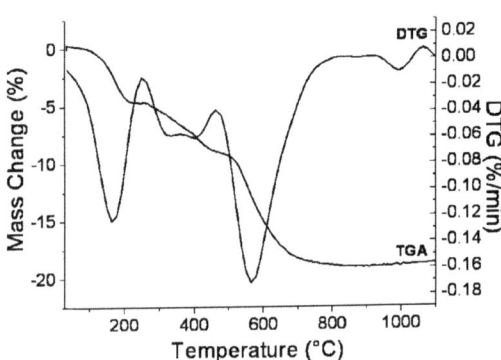

Figure 47: TGA and DTG of Ceraset in argon.

Selected MS spectra of gaseous byproducts relative to the thermal transformation of Ceraset are shown in Figure 48. MS analysis of Ceraset, previously crosslinked at two different temperatures, is present in literature [Li2001].

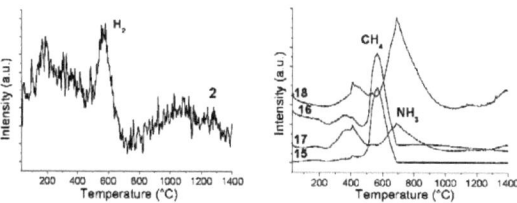

Figure 48: Selected TG/MS analysis of the gaseous byproducts of the thermal decomposition of Ceraset. The numbers in the graphs refer to m/z.

The TGA and DTG graphs of the polysilazane Ceraset display four mass losses. The first weight loss (up to 250 °C) is mainly due to the evaporation of low molecular weight oligomers, which is not detectable by MS analysis, as their weight is too high to be detected. As MS analysis shows, also the elimination of a small amount of hydrogen occurs during the first step. The second and third steps (250-450 °C) are due to H_2 and NH_3 evolution due to the polycondensation reactions (dehydrocoupling and transamination). The forth mass loss, between 450 and 700 °C, is associated with mineralization reactions resulting in the loss of H_2, NH_3 and CH_4.

6.2.1.5.1. Ceraset heat-treated at 500 °C

Among all the Ceraset samples analyzed, the most suitable for LED applications is Ceraset after annealing at 500 °C. Therefore, its thermal stability was proven by means of TGA. The thermal behavior in air and argon of the previously annealed Ceraset at 500 °C is shown in Figure 49.

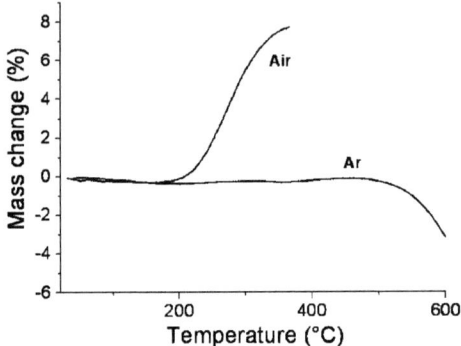

Figure 49: TGA analysis of the Ceraset sample pre-annealed at 500 °C performed in air and argon.

The TG analysis performed in argon of Ceraset previously annealed at 500 °C shows no mass change up to 500 °C; thus, the sample is thermally stable in argon up to the temperature of its first treatment. The same sample analyzed in air displays no mass change only up to 200 °C. At higher temperatures oxidation takes place and the sample increases in weight because of the formation of SiO_2. For LED devices which reach temperatures of 150 °C in air, the thermal stability of Ceraset treated at 500 °C could be sufficient. However, the long term sample stability should be analyzed at this temperature.

6.2.1.6. XRD measurements

The XRD analysis of Ceraset annealed at different temperatures is displayed in Figure 50.

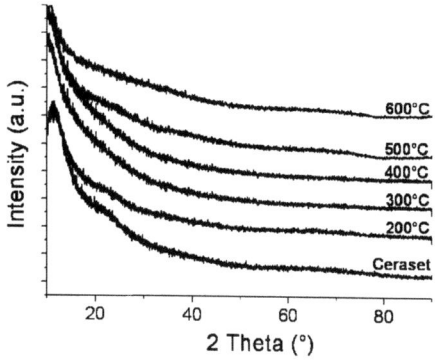

Figure 50: X-ray diffraction patterns of Ceraset samples heat-treated at different temperatures.

Since no crystalline phase is detected with XRD measurements in any of the samples, Ceraset remains amorphous after heat-treatment up to 600 °C.

6.2.1.7. Multinuclear Solid State MAS NMR spectroscopy (^{1}H, ^{13}C and ^{29}Si)

The thermal transformation of the polymer Ceraset to SiCN ceramic studied by solid state NMR was previously reported, as well as the liquid state NMR of the Ceraset precursor [Li2001, Kamperman2007]. In the present thesis, complementary NMR results on the thermolysis of Ceraset are presented [Menapace2008]. It must be considered that the heating rate of the annealing used in this work is inferior compared to that applied in the literature.

^{1}H, ^{13}C and ^{29}Si MAS NMR spectra of Ceraset heat-treated at 200, 300, 400, 500 and 600 °C are shown in Figure 51 and their chemical shifts and bonds attribution are listed in Table 12.

6. Results and discussion

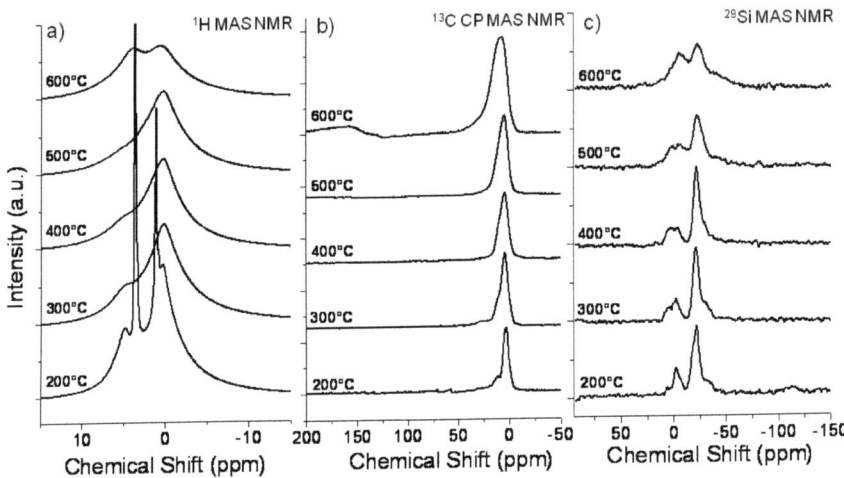

Figure 51: a) ^{1}H, b) ^{13}C CP and c) ^{29}Si MAS NMR spectra of Ceraset heat-treated at different temperatures.

Table 12: Chemical shifts and relative bonds detected in MAS NMR spectra for Ceraset samples heat-treated at different temperatures.

	^{1}H MAS NMR	^{13}C CP MAS NMR	^{29}Si MAS NMR
Ceraset 200 °C	0.3, 0.8, 1.0 ppm (CH_3, CH_2); 3.5 ppm (N-H); 4.8 ppm (Si-H, protonated sp^2 carbon (-HC=))	3.7 ppm (Si-CH$_3$, Si-CH$_2$-CH$_2$-Si); 11.9 ppm (SiCH$_2$Si); 15.2 ppm (CH$_3$-CH$_2$-, CH$_2$-CH$_2$-CH$_2$-); 26.5 ppm (C-C-C); 58.0, 71.3 ppm (N-C); 134.8, 143.3 ppm (=CH$_2$, -CH=)	6.4 ppm (SiC$_3$N); -2.7 ppm (SiC$_3$N, SiC$_2$N$_2$); -21.4 ppm (SiC$_2$N$_2$, SiHCN$_2$)
Ceraset 300 °C	0.2 ppm (CH_3, CH_2); 5.0 ppm (N-H, Si-H, adsorbed water)	4.3 ppm (Si-CH$_3$, Si-CH$_2$-CH$_2$-Si); 10.3 ppm (SiCH$_2$Si); 27.2 ppm (C-C-C)	4.4 ppm (SiC$_3$N); -2.3 ppm (SiC$_3$N, SiC$_2$N$_2$); -21.3 ppm (SiC$_2$N$_2$)
Ceraset 400 °C	0.2 ppm (CH_3, CH_2); 5.0 ppm (N-H, Si-H, adsorbed water)	5.0 ppm (Si-CH$_3$, Si-CH$_2$-CH$_2$-Si); 10.6 ppm (SiCH$_2$Si); 20.5 ppm (C-C-C)	3.7 ppm (SiC$_3$N); -3.6 ppm (SiC$_3$N, SiC$_2$N$_2$); -21.6 ppm (SiC$_2$N$_2$)

Ceraset 500 °C	0.3 ppm (CH_3, CH_2); 5.0 ppm (adsorbed water)	5.7 ppm (Si-CH_3, Si-CH_2-CH_2-Si); 152.7 ppm (Csp^2-N)	2.4 ppm (SiC_3N): -6.3, -22.9 ppm (SiC_2N_2); -42.6 ppm (SiCN_3, SiN$_4$)
Ceraset 600 °C	0.2 ppm (CH_3, CH_2); 4.2 ppm (unsaturated CH_2)	9.9 ppm (Si-CH_3, Si-CH_2-CH_2-Si), 133.6 ppm (Csp^2), 149.9 ppm (Csp^2-N)	-6.8, -24.0 ppm (SiC_2N_2); -42.5 ppm (SiCN_3, SiN$_4$)

6.2.1.7.1. ^{1}H MAS NMR

^{1}H MAS NMR spectra of Ceraset heat-treated at 200, 300, 400, 500 and 600 °C are shown in Figure 51 (a). The sample annealed at 200 °C presents two multiplets, the first corresponding to CH_3 bonded to the silicon (0.3, 0.8, 1.0 ppm), the second to the N-*H* bonding (3.5 ppm) and the overlapping of Si-*H* bonding and protonated sp^2 carbon (-*H*C=) (4.8 ppm). For the samples heat-treated from 300 to 500 °C, the proton NMR displays two broad signals: the first around 5.0 ppm (shoulder), due to the overlapping of N-*H*, Si-*H* bonding and at a smaller extent protonated sp^2 carbon (-*H*C=), the second in the range 0.2-0.3 ppm, due to C-*H* bonding. The broadening of the peaks with increasing temperature indicates the increasing crosslinking of the material, as already reported in literature [Li2001]. The N-*H*/Si-*H*/-*H*C= shoulder decreases in intensity with increasing annealing temperature, due to the crosslinking reactions (hydrosilylation, dehydrocoupling and transamination) and the finding is in accordance with other structural analyses (FT-IR, Raman). Nevertheless, adsorbed water also contributes to this peak. At 600 °C, a new peak centered at 4.2 ppm is observed and assigned to unsaturated CH_2 bonding (sp^2 carbon). Due to the proximity of the two peaks it is not possible to state whether the N-*H*/Si-*H*/-*H*C= peak is still present at 600 °C, having, if present, too low intensity. At 600 °C, vinyl groups are completely absent, while Si-H and N-H bonds are significantly decreased, though still visible in FT-IR and Raman.

6.2.1.7.2. ^{13}C CP MAS NMR

^{13}C CP MAS NMR spectra of Ceraset heat-treated between 200 and 600 °C are shown in Figure 51 (b). With the increase of curing temperatures, the main peak shifts from 3.7 ppm at 200 °C to 9.9 ppm at 600 °C (Si-CH_3) with significant broadening, which is caused by a gradually increasing number of Si-C bonds [Li2001]. At 200 °C the main peak at 3.7 ppm is assigned to Si-CH_3 or Si-CH_2-CH_2-Si bonds, the peak at 11.9 ppm appearing as a shoulder of the main peak is assigned to CH_2 bonded to two silicon atoms, the peak centered at 15.2 ppm is due to CH_3-CH_2- or CH_2-CH_2-CH_2- chains and the peak at 26.5 ppm could be assigned to C-C-C chains. The carbon chains arise from hydrosilylation and vinyl polymerization reactions. Two peaks are visible at 58.0 and 71.3 ppm and can be assigned to N-C bonds, corresponding to the R′ alkyl groups (illustrated in the Ceraset structure) bonded to the nitrogen [Li2001]. Other two peaks centered at 134.8 and 143.3 ppm are present and refer to =CH_2 and –CH= from vinyl groups (–CH=CH_2). At 300 and 400 °C, the main peaks, relative to Si-CH_3 or Si-CH_2-CH_2-Si bonds, are broadened and detectable at 4.3 and 5.0 ppm, respectively. The shoulder at around 10.3 ppm is still visible (SiCH_2Si). C-C-C bonds are present at 27.2 and 20.5 ppm, respectively. The N-C bonds are not visible anymore, as well as the vinyl groups, indicating the occurred crosslinking reactions through vinyl groups. At 500 °C, the main peak is shifted to 5.7 ppm and further broadened and a new very weak peak is detectable at 152.7 ppm (carbon sp^2-N). No C-C-C bonds are detected anymore, probably because they start to convert to sp^2 free carbon. At 600 °C, the main peak is shifted to 9.9 ppm and two types of carbon sp^2 at 133.6 ppm and 149.9 ppm were distinguished, which can be assigned to free carbon sp^2 and to the carbon sp^2 bonded to the nitrogen. The same carbon species were found in our group for carbon-rich SiCN ceramics [Mera2009]. Mera *et al.* assigned the peaks at 124 and 140 ppm to carbon–carbon and carbon–nitrogen double bonds respectively, correlated to the formation of a nitrogen containing sp^2 carbon phase. The same chemical shift was previously discerned for heterocyclic molecules [Balci2005]. Therefore, in the present study, the formation of a network of aromatic carbon atoms containing substitutional nitrogen is plausible.

6.2.1.7.3. ^{29}Si MAS NMR

^{29}Si MAS NMR spectra of Ceraset heat-treated at 200, 300, 400, 500 and 600 °C are shown in Figure 51 (c). The spectrum of Ceraset 200 °C exhibits three resonance peaks centered at 6.4, -2.7 and -21.4 ppm. According to the literature, the resonance at 6.4 ppm corresponds to [SiC$_3$N] end groups, the peak at -2.7 ppm is due to [SiC$_2$N$_2$] or [SiC$_3$N] sites and the peak at around -22 ppm was assigned to CH$_3$HSiN$_2$ ([SiHCN$_2$]) or to [SiC$_2$N$_2$] [Li2001]. At 300 and 400 °C, the -2.7 ppm band broadens significantly (in the range between -4.0 to 7.1 ppm), because of the increased degree of crosslinking [Li2001]. The peak at around -22 ppm is still the main peak. At 500 °C and 600 °C, the intensity ratio between the band centered at -2.7 ppm and the main peak at around -22 ppm increases, denoting the higher degree of crosslinking through evolution of organic groups and the transition from polymer to ceramic. A new broad shoulder centered at -42.6 ppm is detected at 500 °C and increases at 600 °C, assigned to the contributions of [SiCN$_3$] and [SiN$_4$] environments.

6.2.1.7.4. Discussion of the Solid State MAS NMR results

Solid state MAS NMR analysis confirms that Ceraset crosslinks through vinyl groups. At low temperature treatments vinyl groups are detected, while at higher treatment temperatures their amount decreases. C=C double bonds are visible again at 500 °C, due to the formation of sp^2 free carbon. The solid state ^{13}C CP NMR analysis suggests that free carbon starts to form at 500 °C, where it is slightly detectable, and further develops at 600 °C. The present results are in accordance with Raman measurements; however the temperature of free carbon formation is lower than that found in previous studies, probably related to the lower heating rate used during annealing [Li2001]. Moreover, at 600 °C two types of sp^2 carbon atoms can be distinguished in the ^{13}C NMR spectrum, at 133.6 and 149.9 ppm, assigned to free sp^2 carbon and to the sp^2 carbon bonded to nitrogen, respectively. The presence of sp^2-hybridized free carbon is ascertained also from the slightly yellowish color after treatment at 500 °C and the red-brown color of the sample heat-treated at 600 °C. As in the case of MK, the temperature increase results in an enhanced degree of crosslinking of the Ceraset polymer and differently coordinated Si atoms, namely [SiHCN$_2$], [SiC$_2$N$_2$], [SiCN$_3$] and [SiN$_4$]. At 600 °C, the broadening of the peaks observed and the free carbon development, as revealed in proton NMR (protonated carbon sp^2 at 4.2 ppm) and carbon CP NMR (133.6 ppm) denotes the transition from polymer to ceramic.

6.2.1.8. EPR measurements

As in the case of MK, also for Ceraset a correlation between dangling bonds and fluorescence was hypothesized. X-band EPR measurements obtained at 15 K of Ceraset heat-treated at different temperatures (in grey) and their numerical simulations (with Matlab) (in black) are shown in Figure 52.

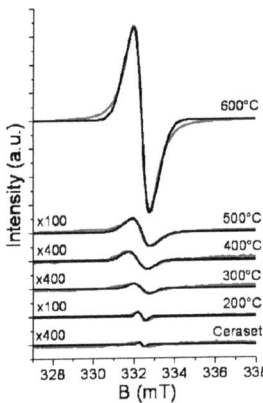

Figure 52: X-band EPR spectra (obtained at 15 K) of Ceraset samples heat-treated at different temperatures (in grey) with simulations (in black). The signals relative to the low temperature treated samples were magnified by a factor (labeled) for representative reasons.

The EPR spectra are characteristic for dangling bonds. The quantitative analysis of the X-band EPR signals simulations provides information about g-factor, line width and spin concentration of the radicals (Table 13).

6. Results and discussion

Table 13 Quantitative analysis of X-band EPR signals simulations for Ceraset samples.

Sample	g-factor	Line width (mT)	Concentration (spin·mg^{-1})
Ceraset	2.001	0.25	3.20·10^8
Ceraset 200 °C	2.001	0.39 / 1	8.43·10^9
Ceraset 300 °C	2.001	0.9 / 1.6	3.69·10^9
Ceraset 400 °C	2.002	1.1	7.20·10^9
Ceraset 500 °C	2.001	0.7 / 1.35	5.91·10^{10}
Ceraset 600 °C	2.001	0.65 / 1.35	3.30·10^{13}

As in MK samples, the *g*-values are comprised between 2.001 and 2.002, which are typical for the carbon species, in accordance with the results previously obtained for similar materials [Andronenko2006, Trassl2002, Berger2005].

Almost all Ceraset samples consist of two overlapping resonances with identical *g*-value but different line widths. The values of the line widths for different treatment temperatures are represented in Figure 53.

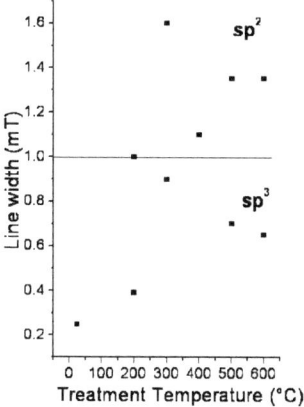

Figure 53: Line widths of the resonances of the Ceraset samples.

The sp^3-type carbon dangling bonds are observed in all Ceraset samples except 400 °C, while sp^2-hybridized species are found at 200, 300, 400, 500 and 600 °C. The EPR analysis of liquid Ceraset indicates the absence of sp^2 carbon radicals. Therefore, there are no radicals sited on vinyl groups, which are unreacted and perfectly stable. After treatment at 200 and 300 °C, the vinyl groups start to react and sp^2 carbon radicals, sited on vinyl groups, are detectable. The sample heat-treated at 400 °C shows the presence of a single signal with line width corresponding to sp^2 carbon radicals. 400 °C is a temperature for which C=C double bonds from vinyl groups should be significantly reduced, while C=C bonds from free carbon are not yet formed. The samples heat-treated at 500 and 600 °C show the presence of both sp^2 and sp^3 carbon radicals. In this case, sp^2 carbon radicals are the precursor of the free carbon phase.

Because of the presence of vinyl groups in the starting polymer, and consequently the detection of sp^2-type carbon radicals at low annealing temperatures, in the case of Ceraset the recognition of the temperature of free carbon formation with EPR analysis is not straightforward, as in the case of MK. Nevertheless, Raman, solid state MAS NMR and optical observations indicate the presence of free carbon at a temperature as low as 500 °C.

The spin concentrations of the Ceraset samples are illustrated in function of the treatment temperature in Figure 54 (the y-axis is in decimal logarithmic scale). Their exponential fit (in base 10) is also shown (linear fit in logarithmic scale).

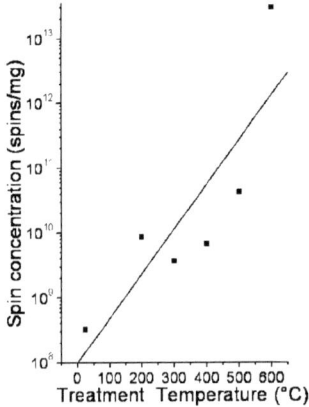

Figure 54: Concentration of dangling bonds in Ceraset samples in function of the treatment temperature and fit of the exponential growth. Notice that the y-axis is in decimal logarithmic scale.

The radical concentration increases with the heat-treatment temperature. As for MK, an exponential growth can be extrapolated (SD=1, R=0.85). The heat-treatment of the polysilazane at 600 °C results in a spin concentration which is 3 to 4 orders of magnitude higher than that of the other samples. Similar values were found also for MK at 700 °C, thus the higher radicals concentration should be related to the higher bond cleavage activity related to the polymer-ceramic transition in comparison to lower temperatures. If the spin concentration corresponding to the Ceraset annealed at 600 °C is excluded from the graph, a better linear fit is extrapolated (SD = 0.40, R = 0.89).

The spin concentrations of the Ceraset samples plotted toward their fluorescence intensities, assumed as the integral of the fluorescence spectra in energy scale, are illustrated in Figure 55. Ceraset 600 °C is not shown for representative reasons.

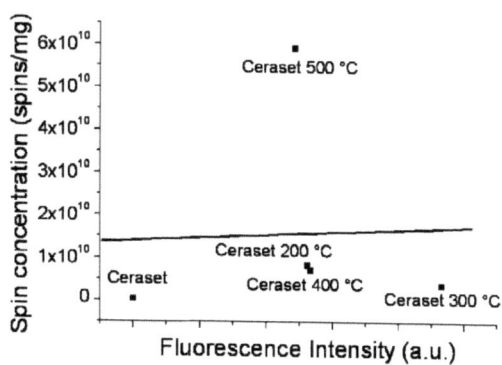

Figure 55: Correlation between spin concentration and fluorescence intensity of the Ceraset samples.

As in the case of MK polymer, no linear correlation can be detected. Dangling bonds, although present in the samples, are not responsible for the photoluminescence properties, although it is not excluded that they could have a marginal role.

6.2.1.9. Discussion

The pyrolysis behavior of the polysilazane Ceraset is very similar to that relative to the polysiloxane MK. Many correspondences were found, although the structural changes were located at different temperatures. For example, the free carbon formation was detected by solid state MAS NMR in MK heat-treated at 700 °C (in proton NMR at 600 °C) and in Ceraset annealed at 500 °C, although clearly visible only at 600 °C (Figures 22 and 51). From EPR measurements, sp^2-hybridized carbon radicals were found at lower temperatures, at 500 °C for both MK and Ceraset (Figures 24 and 53). For Ceraset, an additional confirmation of the presence of sp^2 carbon at 500 °C was provided by Raman spectra (Figure 46). Optical observations are also very important to detect free carbon, as the color indicates its concentration. The samples MK 700 °C and Ceraset 600 °C are brown, while at 500 °C both annealed polymers appear yellow. The differences in the detection of sp^2 carbon, by means of MAS NMR, EPR and Raman measurements could be assigned to the lower sensitivity of solid state NMR compared to EPR and Raman analyses. Moreover, the resolution of solid state MAS NMR spectra is considerably reduced for samples containing dangling bonds.

In the view of the parallelism between MK and Ceraset concerning the formation of free carbon, the considerations elucidated for MK on the formation of aromatic agglomerations, and their absorption and photoluminescence properties, are also valid for Ceraset. Moreover, unlike MK, the Raman spectra of Ceraset 500 and 600 °C show the presence of C=C bonds, and this could indicate a higher order in the sp^2 carbon structures in Ceraset compared to MK.

Ceraset is the first polymer analyzed in this thesis that highlights the relationship between crosslinking and photoluminescence properties. In the precursor state, Ceraset was not crosslinked and showed weak fluorescence (Figure 33). On the contrary, MK was highly crosslinked in the initial state and showed an appreciable luminescence in the UV. If crosslinking has a role in fluorescence, as we hypothesize, the fluorescence of MK depends on its highly crosslinked state. The weak fluorescence emission of liquid Ceraset could arise from luminescent species created by the slow, natural crosslinking occurring during shelf life (aging). When partially crosslinked at 200 °C, Ceraset shows an appreciable emission (Figure 32). After heat-treatment at 300 °C, the crosslinking through vinyl groups can be considered complete and the fluorescence intensity results improved. The excitation and emission spectra relative to Ceraset 200 and 300 °C have the same shape with the same maximum (319 nm). Consequently, the luminescent species that origins the emission should also be the same. The only structural change occurring between 200 and 300 °C is the crosslinking extent, evident in solid state NMR spectra (Figure 51). Therefore, higher crosslinking extent implies higher emission intensity, correlated to the formation of the luminescent species. Up to 300 °C a single crosslinking mechanism dominates, i.e. the hydrosilylation [Li2001]. Therefore, the luminescent species formed through hydrosilylation could be responsible for the emission spectra with maximum at 319 nm. In the case of MK, different emission peaks are supposed to identify the presence of different emitters, and their intensity ratio changes as the thermal crosslinking proceeds. In Ceraset, up to 300 °C a single emitter was observed, probably formed by the hydrosilylation reaction. At annealing temperatures higher than 350 °C, also vinyl polymerization, dehydrocoupling and transamination reactions occur [Li2001]. The new crosslinking reactions generate new luminescent species, as attested by the emission spectrum of Ceraset after treatment at 400 °C, which is red-shifted compared to those relative to the samples heat-treated at 200 and 300 °C. This means that the new luminescent species built are characterized by lower band gap. An increased contribution of the low energy bands was observed in both excitation and emission spectra. After annealing at 500 °C a further red-shift is observed. At this temperature the contribution of crosslinking to photoluminescence properties is not the only one, as free carbon starts to be detected. After treatment at 600 °C, further red-shift and decrease of the fluorescence intensity are observed, due to the further development of free carbon. The photoluminescent species which develop with the crosslinking have not yet been clarified.

As in MK, the role of dangling bonds was not highlighted. If they have an influence on the photoluminescence properties, this is marginal. Therefore, the silicon-based polymers successively analyzed in this thesis were not subjected to EPR analysis.

6.2.2. KiON VL20

KiON Ceraset Polysilazane 20 or VL20 is a commercial polysilazane very similar to the KiON Ceraset Polyureasilazane, with the difference that Ceraset additionally contains urea functionality. Also VL20 contains cyclic and linear chains, but fewer low molecular weight polysilazane components [KiON].

In the same way as Ceraset, VL20 crosslinks to a rigid and insoluble transparent solid by heating at 180-200 °C or at lower temperatures by adding a free radical initiator (organic peroxide) or by exposure to UV radiation in the presence of a UV sensitizer [KiON].

Similarly to Ceraset, VL20 is pyrolyzed to silicon carbide, silicon nitride or SiCN ceramics at elevated temperatures in the presence of argon, nitrogen or ammonia. VL20 is suitable for the same applications of Ceraset, such as MEMS, composites, porous ceramics, materials for electrochemical applications [Asthana2006, Jones2009, Lee2007, Sarkar2008, Cramer2004, Gadow2003, Kolb2006, Liebau-Kunzmann2006, Lee2005, Nghiem2007, Pham2006, Lee2006, Reddy2004, Roh2008, Reddy2003, Stantschev2005, Wilden2007].

The structure provided by the Kion Company is illustrated in Figure 56.

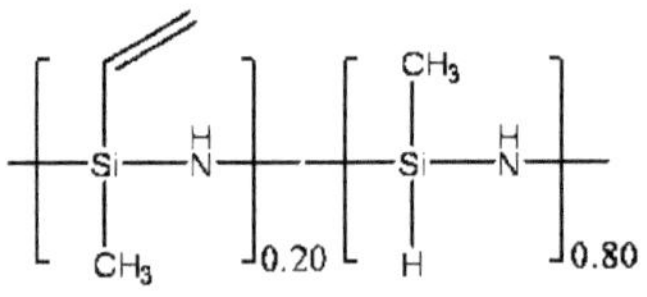

Figure 56: Molecular structural units of KiON Ceraset Polysilazane 20 as provided by KiON Corp.

As in the case of Ceraset, the polymer possesses crosslinkable substituents (vinyl, Si-H and N-H groups) and it is composed by small molecular weight oligomers (although fewer) as well as high molecular weight polymer chains.

The liquid VL20 was heat-treated at 200, 300, 400, 500 and 600 °C for 2 hours under Ar flow, with heating rate 50 °C/h and free cooling. As in the case of Ceraset, solid crosslinked samples were obtained, which retained the shape of the mold, and were subsequently powdered.

6.2.2.1. Photoluminescence measurements

6.2.2.1.1. Fluorescence measurements

Photoluminescence properties were observed for all heat-treated samples. Figure 57 displays the excitation and emission spectra of the VL20 annealed at different temperatures. The excitation and emission wavelengths used for the measurements correspond to the maximum intensity peaks (Table 14).

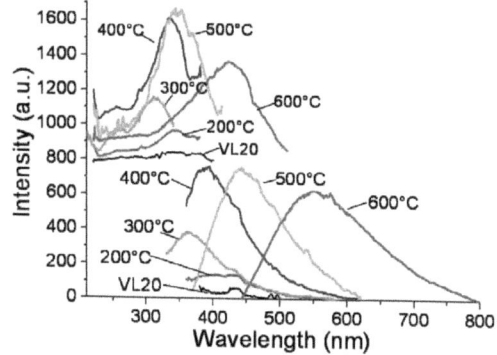

Figure 57: Fluorescence emission (bottom right) and excitation (top left) spectra of VL20 samples obtained with excitation and emission wavelengths according to Table 14.

Table 14: Excitation and emission wavelengths used for the polysilazane VL20 samples.

Sample	Excitation wavelength (nm)	Emission wavelength (nm)
VL20	360	433
VL20 200 °C	340	400
VL20 300 °C	300	363
VL20 400 °C	329	394
VL20 500 °C	349	436
VL20 600 °C	420	527

Blue and blue-greenish emission was obtained for samples heat-treated up to 500 °C, with maximum intensities for heat-treatments at 400 and 500 °C.

The untreated polymer VL20 shows weak luminescence in the UV/blue range. After heat-treatment at low temperatures (200, 300 °C) the polymer shows increasingly intense photoluminescence emissions. Maximum excitation peaks were detected at 360, 340 and 300 nm and maximum emission peaks at 433, 400 and 363 nm, for the untreated VL20 and VL20 annealed at 200 and 300 °C, respectively. Nevertheless, up to 200 °C the fluorescence emission is very weak and comparable to the scattering of the laser beam inside the sample compartment.

After annealing at 400 °C, beyond the increased emission intensity, a red-shift of the emission maximum in respect to the sample annealed at 300 °C is also appreciable. The maximum peak is now at 394 nm with excitation maximum at 329 nm.

After heat-treatment at 500 °C, a further red-shift of the fluorescence emission peak was detected, while the intensity does not vary appreciably. The maximum excitation peak was found at 349 nm, while the maximum emission peak appears at 436 nm.

After annealing at 600 °C, an ulterior red-shift is detectable: the excitation maximum is at 420 nm and the emission maximum at 527 nm. Additionally, the sample exhibits lower photoluminescence intensity in comparison to the previous sample (calculated as the integral under the spectrum in energy scale), evident also from the q.e. results reported later and the optical observations. The intensity decrease at this annealing temperature is due to the development of free carbon.

VL20 samples annealed at different temperatures are shown in Figure 58 under white (a) and UV (360-400 nm) (b) light.

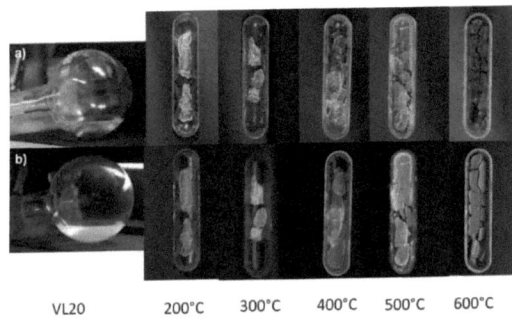

Figure 58: Photographs representing VL20 samples under white (a) and UV (b) light. For colored pictures the reader can refer to http://tuprints.ulb.tu-darmstadt.de/2085/.

As in the case of MK polymer and Ceraset, the samples are transparent and colorless up to 400 °C. The sample treated at 500 °C acquires a yellow coloration remaining transparent and shows visible luminescence if irradiated with UV light; at 600 °C the material turns translucent red-brown under white light and opaque under UV light.

In particular, the optical and photoluminescence properties of VL20 resemble the ones obtained for Ceraset: similarities are the red-shift of the emission and excitation peaks as the temperature increases from 300 to 600 °C, the similar emission ranges for the same annealing temperatures, the yellow and red-brown colors of the samples heat-treated at 500 and 600 °C respectively, and the decrease of the emission intensity at 600 °C.

As in the case of Ceraset, the heat-treatment of VL20 at 500 °C results in an interesting emission in the visible range with excitation maximum at around 360 nm. The color of the annealed sample is very similar to that of Ceraset at 500 °C, and similarly transparent. VL20 annealed at 400 °C is also an interesting sample, transparent with emission in the blue range.

In Figure 59 the emission spectra of the VL20 obtained with excitation wavelength of 250 nm (a) and 360 nm (b) are shown.

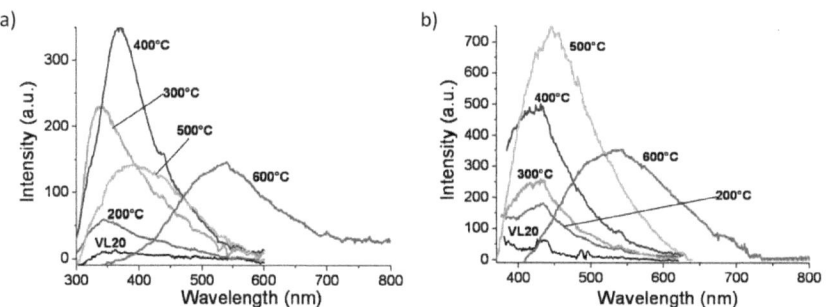

Figure 59: Fluorescence emission spectra of VL20 annealed at different temperatures obtained with excitation wavelength 250 nm (a) and 360 nm (b).

The sample annealed at 400 °C exhibits the most intense emission with excitation wavelength 250 nm, while for 360 nm excitation the sample annealed at 500 °C displays the strongest emission. This is in accordance with the excitation spectra in Figure 57, where two peaks at around 250 nm and 360 nm are detectable in the samples heat-treated at 400 and 500 °C, the first being more intense in VL20 400 °C, the second in VL20 500 °C. For both excitation wavelengths, the fluorescence intensity decreases at 600 °C.

The spectral emissions obtained with 360 nm excitation of the VL20 samples heat-treated at 200-600 °C are represented on the CIE chromaticity diagram 1931 (Figure 60).

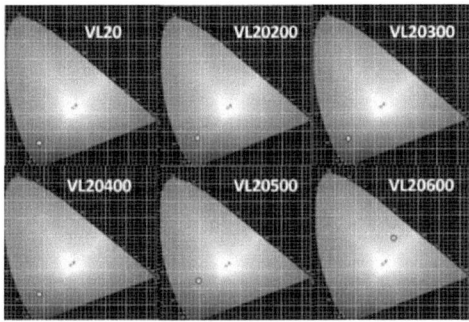

Figure 60: CIE chromaticity diagrams 1931 of the VL20 samples annealed at 200-600 °C. For colored pictures the reader can refer to http://tuprints.ulb.tu-darmstadt.de/2085/.

The color coordinates are in accordance with the colors of the samples under UV light (Figure 58).

6.2.2.1.2. Stability of the fluorescence during storage

In order to monitor the change in fluorescence properties of the VL20 samples following storage in air, the fluorescence emission spectrum obtained with 250 nm excitation wavelength was detected in the first week after the heat-treatment and about one year later (Figure 61).

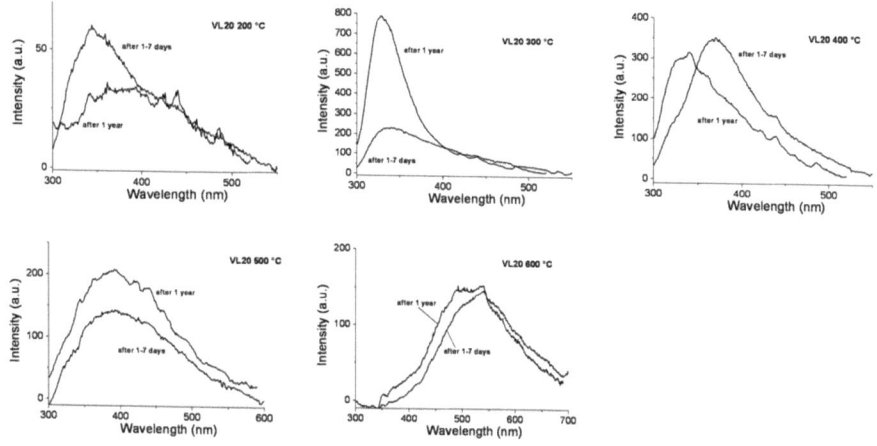

Figure 61: Fluorescence spectra of the VL20 samples obtained during the first week after heat-treatment and 1 year later (at 250 nm excitation).

The fluorescence of the VL20 sample annealed at 200 °C slightly decreases in time; the intensity relative to the sample treated at 300 °C significantly increases; for the sample treated at 400 °C the emission maximum is blue-shifted and for VL20 500 and 600 °C the fluorescence intensity changes only slightly. For low temperature treatments (200-400 °C) hydrolysis reactions with moisture probably occur, while for high temperature treatments (500-600 °C) the samples are more stable in air and smaller changes in fluorescence are detected.

6.2.2.1.3. Quantum efficiency

The quantum efficiencies of the VL20 samples heat-treated at 400 and 600 °C were measured with the integrating sphere, resulting to be 11% and 3.4%, respectively. The quantum efficiencies of the remaining VL20 samples were estimated by comparing the integral of their emission spectra (in energy scale) with the integral of the emission spectra of samples of known quantum efficiency with emission in the same range (Table 15).

Table 15: Quantum efficiencies of the VL20 samples.

Sample	Quantum efficiency (%)
VL20	0.5
VL20 200 °C	3.1
VL20 300 °C	6.2*
VL20 400 °C	11.0
VL20 500 °C	6.3
VL20 600 °C	3.4

The samples labeled with * emit in the UV range (<394 nm), for which no measurement was successful; therefore the values were calculated on comparison with a sample with emission maximum at 394 nm.

6.2.2.2. Absorption measurements

Additional information about the electronic transitions occurring in the VL20 samples and their variation after temperature treatment can be obtained with absorption measurements.

6.2.2.2.1. UV-Vis-NIR spectroscopy

As already attested for MK polymer and Ceraset, the solubility of VL20 decreases as the treatment temperature increases, and standard UV-Vis measurements could be only performed on the polysilazane VL20 and on the heat-treated sample at 200 °C, in THF solution (Figure 62). UV-Vis measurements were previously carried out on VL20 [Pham2006] and on VL20 samples heat-treated at low temperatures (up to 300 °C) in nitrogen and oxygen atmospheres, the last ones in solid state [Asthana2006].

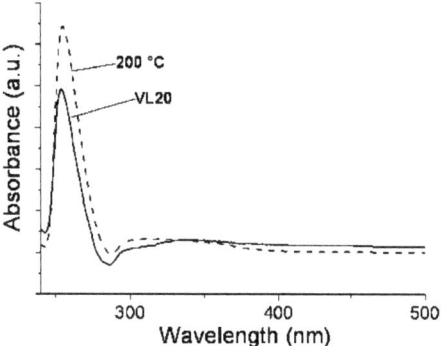

Figure 62: UV-Vis absorption spectra of VL20 (solid line) and of the polysilazane heat-treated at 200 °C (dashed line).

The absorption spectrum of VL20 annealed at 200 °C is almost coincident with that of VL20, when normalized. The absorption edge occurs in both cases at around 285 nm (4.35 eV). Pham *et al.* found that the absorption edge of VL20 was at 225 nm (5.51 eV), although not specifying the solvent [Pham2006]. Asthana *et al.* characterized two VL20 samples with dicumyl peroxide at different crosslinking stages, both in solid state, and showed the decrease in transmittance in the visible range in the sample further crosslinked in oxygen [Asthana2006]. In the KiON description of the product for the cured resin the absorption edge is indicated at around 300 nm (4.13 eV).

6.2.2.2.2. Reflection measurements

The reflection spectra were obtained using synchronous excitation and detection wavelengths, with the Cary Eclipse Varian. The absorption/scattering spectra obtained from the reflection data relative to the VL20 samples (using the Kubelka-Munk function) are displayed in Figure 63. The spectra were smoothed for representative reasons. The sample VL20 200 °C was also normalized, as its intensity was significantly higher than those of the rest of the samples (due to its transparency related to its gel state).

6. Results and discussion

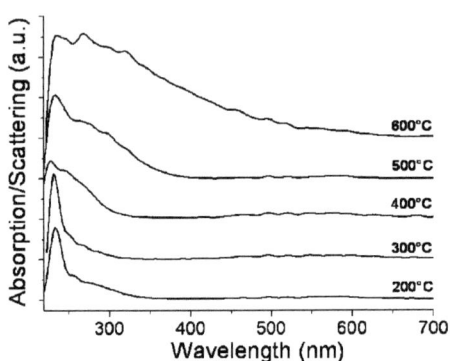

Figure 63: Absorption/scattering spectra relative to the VL20 samples obtained with reflection measurements.

Except for a small absorption band at around 300 nm in the sample heat-treated at 200 °C, the absorption edges of the VL20 samples are shifted toward less energetic wavelengths as the treatment temperature increases. The results are analogous to those relative to Ceraset.

The sample annealed at 300 °C absorbs radiation up to about 320 nm (3.87 eV), VL20 400 °C up to 340 nm (3.65 eV) with increased importance of the absorption at less energetic wavelengths. After heat-treatment at 500 °C, the absorption edge is at about 400 nm (3.10 eV), while for VL20 600 °C at 630 nm (1.97 eV). The absorption spectra are in agreement with the excitation spectra of Figure 57 and with optical observations. Moreover, the results show the same trend of photoluminescence emissions. The sample annealed at 200 °C shows an unusual behavior, which could be attributed to its partially liquid structure, in analogous way as Ceraset.

6.2.2.3. FT-IR spectroscopy

The VL20 and the powdered samples heat-treated at 200, 300, 400, 500 and 600 °C were studied with FT-IR structural analysis (Figure 64). In Table 16 the vibration bands and respective bonds are listed. All samples were measured using the ATR device. For wavenumbers lower than 600 cm^{-1}, the ATR device is not reliable; therefore those values are not represented. FT-IR analysis on VL20 crosslinked in N_2 up to 275 °C was previously published [Asthana2006].

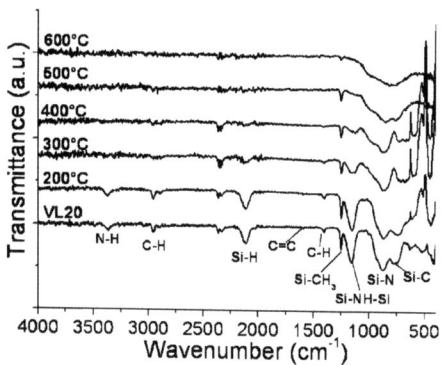

Figure 64: FT-IR analysis of the VL20 samples.

Table 16: Vibration bands and respective bonds detected by means of FT-IR analysis of the VL20 samples.

Bond assignment	Vibration band (cm^{-1})
$\nu(N-H)$	3364
$\nu_{as}(-CH_3)$	2958
$\nu_{as}(CH_2)$	2902
$\nu(C=C)$	1593
$\delta_{as}(C-H)$	1404
$\nu(Si-H)$	2111
$\delta_s(Si-CH_3)$	1253
$\nu(Si-NH-Si)$	1160
$\nu_{as}(Si-N)$	874
$\delta_s(Si-C)$	780

At 200 °C, the Si-H and N-H bonds are unreacted, while the C=C bonds from the vinyl groups, slightly visible in the precursor, have disappeared. At 300 °C, the intensity of the Si-H peak is strongly reduced, while the N-H and -CH$_3$ peaks are not visible anymore. At 400 °C, the intensity of the Si-H peak is further reduced. At 500 °C, the Si-H peak disappears, while the peaks relative to Si-N and Si-C bonds broaden. At 600 °C, the Si-N and Si-C peaks further broaden. By means of FT-IR analysis, at a temperature as low as 300 °C, the crosslinking reactions via vinyl groups (hydrosilylation ad vinyl polymerization), Si-H groups (hydrosilylation and dehydrocoupling) and N-H groups (dehydrocoupling and transamination) start to be detected. At 500-600 °C the ceramic network starts to form. The Si-CH$_3$ and the Si-NH-Si bands decrease in intensity as the treatment temperature increases, indicating the mineralization of the polymer and the occurred crosslinking via dehydrocoupling and transamination, respectively.

6.2.2.4. Raman spectroscopy

The VL20 samples were investigated with confocal micro-Raman spectrometer (Figure 65 and Table 17). An excitation laser of 488 nm was found optimal in order to minimize the fluorescence background. The Raman spectrum of the VL20 precursor obtained at 1064 nm laser wavelength was reported also in the literature [Liebau-Kunzmann2006].

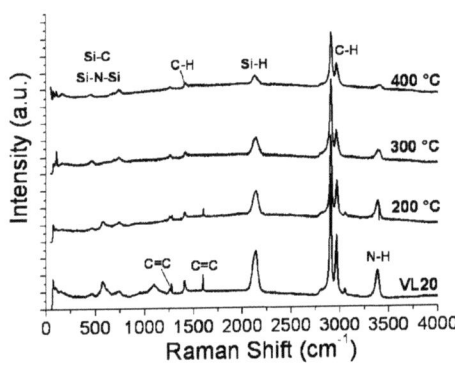

Figure 65: Raman spectra of the VL20 samples using 488 nm as laser wavelength.

Table 17: Vibration bands and respective bonds detected by means of Raman analysis of the VL20 samples.

Bond assignment	Raman shift (cm^{-1})
$v(N-H)$	3384-3397
$v_{as}(-CH_3)$	2960
$v_s(CH_2)$	2902
$\delta(C-H)$	1406-1412
$v(Si-H)$	2136
$v(C=C)$	1595
$v_s(Si-C)+v(Si-N-Si)$	156-945
$v(Si-N)$	474

The C=C bonds from the vinyl groups, visible in VL20 and VL20 200 °C, disappear at higher treatment temperatures, completely reacted via hydrosilylation. The bands relative to the N-H, Si-H and C-H groups as well as the Si-N-Si and Si-C bonds can be observed up to 400 °C. For heat-treatments from 500 °C, the fluorescence of the samples interferes with the Raman measurements and the bands are partially (500 °C) or totally (600 °C) hidden (not shown).
Therefore, the samples heat-treated at higher temperatures (400-600 °C) were also analyzed with the IR/Raman spectrometer Bruker IFS 55 - FRA 106, with laser wavelength 1064 nm. The spectra are reported in Figure 66.

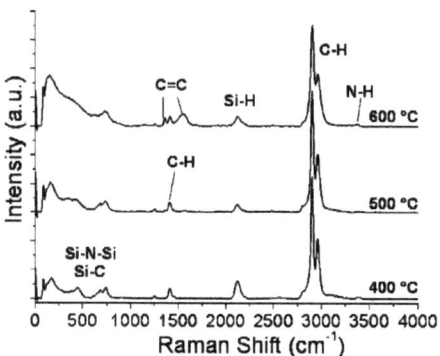

Figure 66: Raman spectra of the VL20 samples heat-treated at 400-600 °C using 1064 nm as laser wavelength.

With laser wavelength 1064 nm it is possible to avoid the luminescence interference. The N-H (3384 cm^{-1}), CH$_2$ (2900 cm^{-1}), -CH$_3$ (2959 cm^{-1}), Si-H (2119 cm^{-1}) and C-H bonds (1410 cm^{-1}) as well as the Si-N-Si and Si-C bonds (up to 1000 cm^{-1}) are now clearly detectable up to 600 °C. Two new bands with maxima at 1358 and 1551 cm^{-1} appear at 500 °C and increase in intensity at 600 °C (Figure 67).

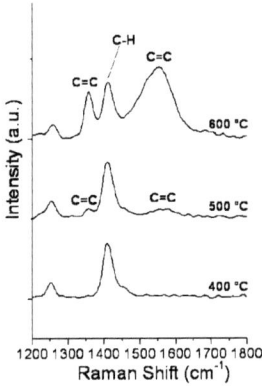

Figure 67: Raman spectra in the range 1200-1800 cm^{-1} of the VL20 samples heat-treated at 400, 500 and 600 °C using 1064 nm as laser wavelength.

Exactly as in the case of Ceraset, the two new bands denote C=C double bonds relative to aromatic agglomerations, although they should not be confused with D and G bands. The formation of sp^2 free carbon detected by Raman analysis is in accordance with optical observations.

6.2.2.5. TGA/DTG /MS

The thermal behavior of VL20 was monitored with STA (Simultaneous Thermal Analysis). TGA- and its differential DTG-curves are represented in Figure 68. TGA analyses of VL20 are present in the literature under argon [Liebau-Kunzmann2006] and N$_2$-Ar conditions [Lee2005].

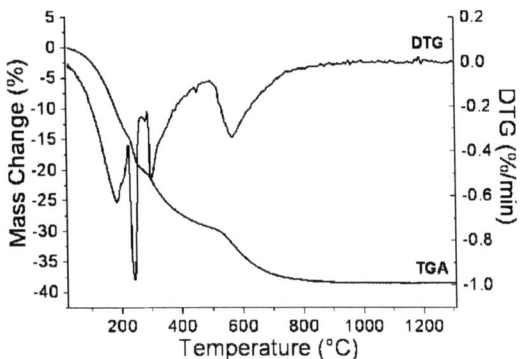

Figure 68: TGA and DTG of VL20 in argon.

Selected MS spectra of gaseous byproducts relative to the thermal transformation of VL20 are shown in Figure 69.

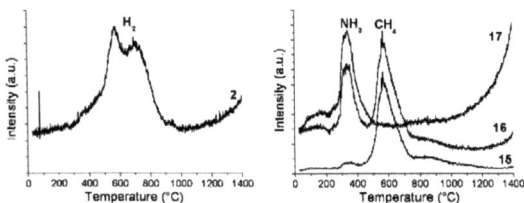

Figure 69: Selected TG/MS analysis of the gaseous byproducts of the thermal decomposition of VL20. The numbers in the graphs refer to m/z.

The TGA of VL20 shows four mass losses: the first, up to 220 °C, is due to oligomers losses, the second and third, between 220 and 450 °C, is due to the elimination of NH_3 (m/z=17, fragment 16), the fourth around 600 °C is due to CH_4 (m/z=15, 16) and H_2 elimination. Oligomers evolution is not detectable by MS analysis, as their weight is too high.

6.2.2.6. XRD measurements

The XRD analysis of the VL20 heat-treated at different temperatures is displayed in Figure 70.

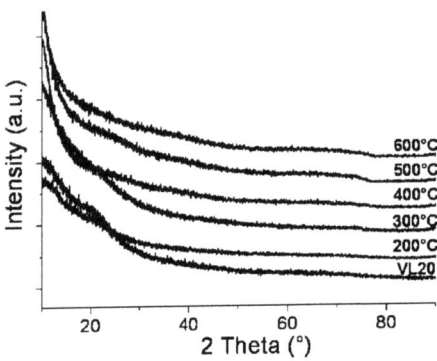

Figure 70: X-ray diffraction patterns of VL20 samples heat-treated at different temperatures.

Since no crystalline phase is detected with XRD measurements in any of the samples, VL20 polymer remains amorphous after heat-treatment up to 600 °C.

6.2.2.7. Multinuclear Solid State MAS NMR Spectroscopy (^{1}H, ^{13}C and ^{29}Si)

^{1}H, ^{13}C CP and ^{29}Si MAS NMR spectra of VL20 heat-treated at 200, 300, 400, 500 and 600 °C are shown in Figure 71 and their chemical shifts and bonds attribution are listed in Table 18.

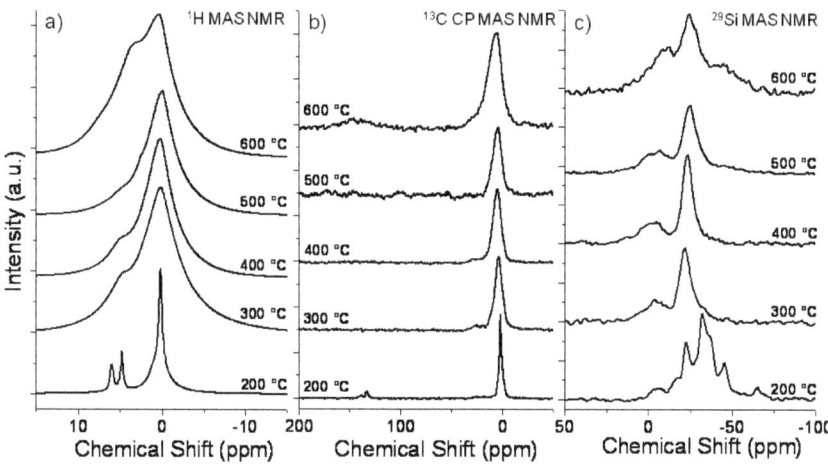

Figure 71: a) ^{1}H, b) ^{13}C CP and c) ^{29}Si MAS NMR spectra of VL20 heat-treated at different temperatures.

Table 18: Chemical shifts and relative bonds detected in MAS NMR spectra for VL20 samples heat-treated at different temperatures.

	^{1}H MAS NMR	^{13}C CP MAS NMR	^{29}Si MAS NMR
VL20 200 °C	0.2, 0.4 ppm (CH_3, CH_2); 4.8, 6.0 ppm (N-H, Si-H, protonated sp^2 carbon (-HC=))	-0.5, 1.9 ppm (Si-CH_3, Si-CH_2-CH_2-Si); 18.8 ppm (CH_3-CH_2-, CH_2-CH_2-CH_2-); 26.4 ppm (C-C-C); 133.3, 139.5 ppm (=CH_2, –CH=)	-65.2, -45.5, -37.4, -32.5 ppm (Si-O); -22.8, -15.8, -5.1 ppm (SiC_2N_2, SiHCN$_2$)
VL20 300 °C	0.2 ppm (CH_3, CH_2); 5.1 ppm (N-H, Si-H, adsorbed water)	1.5, 4.7 ppm (Si-CH_3, Si-CH_2-CH_2-Si); 26.6 ppm (C-C-C)	4.6 ppm (SiC_3N); -4.2 ppm (SiC_3N, SiC_2N_2); -21.8 ppm (SiC_2N_2); -28.1 ppm (SiC_2N_2, SiCN$_3$)
VL20 400 °C	0.4 ppm (CH_3, CH_2); 4.8 ppm (N-H, Si-H, adsorbed water)	3.7 ppm (Si-CH_3, Si-CH_2-CH_2-Si); 28.6 ppm (C-C-C)	-1.3 ppm (SiC_3N, SiC_2N_2); -23.5 ppm (SiC_2N_2)
VL20 500 °C	0.1 ppm (CH_3, CH_2);	3.2, 6.9 ppm (Si-CH_3,	0.3 ppm (SiC_3N,

	2.7 ppm (CH$_2$CH_3); 5.0 ppm (N-H, Si-H, adsorbed water)	Si-CH_2-CH_2-Si); 10.9 ppm (CH_2)	(SiC$_2$N$_2$); -7.1, -24.9 ppm (SiC$_2$N$_2$); -47.6 ppm (SiCN$_3$, SiN$_4$)
VL20 600 °C	0.4 ppm (CH_3, CH_2); 3.7 ppm (unsaturated CH_2); 6.6 ppm (protonated carbon sp^2 (-HC$_{arom}$))	3.8, 8.4 ppm (Si-CH_3, Si-CH_2-CH_2-Si), 13.3 ppm (CH_2); 144.4 ppm (Csp^2)	-8.5, -25.2 ppm (SiC$_2$N$_2$), -45.4 ppm (SiCN$_3$, SiN$_4$)

6.2.2.7.1. ^{1}H MAS NMR

The ^{1}H MAS NMR spectrum of VL20 after heat-treatment at 200 °C shows the presence of CH_3 and CH_2 bonds at around 0 ppm, N-H, Si-H and protonated sp^2 carbon (-HC=) at 4.8 and 6.0 ppm. From 300 to 500 °C, CH_3 and CH_2 bonds are visible; according to ^{13}C CP spectra, FT-IR and Raman, at these temperatures vinyl groups already reacted via hydrosilylation and dehydrocoupling, and protonated sp^2 carbon (-HC=) should have disappeared from the ^{1}H spectrum. A peak at around 5 ppm appears, however due to the poor resolution of the ^{1}H spectra it can be assigned both to adsorption of water as well as to N-H and Si-H groups. At 500 °C, a peak emerges at 2.7 ppm, relative to hydrogen bonded to carbon not directly bonded to a silicon, for example CH$_2$CH_3. At 600 °C, CH_3, CH_2 and unsaturated CH_2 are detected, with the addition of a new signal at 6.6 ppm, indicating protonated carbon sp^2 (-HC$_{arom}$), which is related with the formation of free carbon.

6.2.2.7.2. ^{13}C CP MAS NMR

The ^{13}C CP MAS NMR spectrum of VL20 heat-treated at 200 °C displays the existence of Si-CH$_3$ or Si-CH$_2$-CH$_2$-Si bonds at -0.5 and 1.9 ppm, CH$_3$-CH$_2$- or CH$_2$-CH$_2$-CH$_2$- at 18.8 ppm and C-C-C at 26.4 ppm, arising from hydrosilylation and vinyl polymerization reactions, and unreacted vinyl groups (=CH$_2$ and –CH=) at 133.3 and 139.5 ppm. At 300 and 400 °C, the peaks relative to Si-CH$_3$ or Si-CH$_2$-CH$_2$-Si bonds, detectable at around 0 to 5 ppm, are broadened, and C-C-C bonds are present at 26 to 28 ppm. No vinyl groups are detectable anymore, which suggests that the crosslinking through vinyl groups is completed. At 500 °C the peak relative to Si-CH_3 or Si-CH_2-CH_2-Si is further broadened. CH_2 groups are detected at 10.9 ppm, but no C-C-C bonds are detected anymore, probably because they start to convert to sp^2 free carbon. At 600 °C further broadening of the peak relative to Si-CH_3 or Si-CH_2-CH_2-Si is detected, CH_2 groups are discernible at 13.3 ppm and a new broad peak is detected at 144.4 ppm, indicating free carbon formation (Csp^2).

6.2.2.7.3. ^{29}Si MAS NMR

The ^{29}Si MAS NMR spectrum of VL20 heat-treated at 200 °C shows the presence of several peaks (-65.2, -45.5, -37.4, -32.5 ppm) that indicate the Si-O bond, which denotes the oxidation of the sample, due to the long time of storage in air. Peaks at -22.8, -15.8, -5.1 ppm identify the presence of [SiC_2N_2] and [SiHCN$_2$] bonds. At 300 °C, the peak at 4.6 ppm could indicate the existence of [SiC_3N] terminal groups, while the peaks at -4.2 and -21.8 ppm are related to [SiC_2N_2] units and at -28 ppm to [SiCN$_3$]. At 400 °C, [SiC_3N] terminal groups are detected at -1.3 ppm and [SiC_2N_2] units at -23.5 ppm. At 500 °C, [SiC_3N] terminal groups are detected at 0.3 ppm and more coordination possibilities of the silicon arise, [SiC_2N_2] at -7.1 and -24.9 ppm, [SiCN$_3$] and [SiN$_4$] at -47.6 ppm. At 600 °C [SiC_2N_2], [SiCN$_3$], [SiN$_4$] are present (-8.5, -25.2, -45.4 ppm).

6.2.2.7.4. Discussion of the Solid State MAS NMR results

From ^{1}H and ^{13}C MAS NMR, the free carbon starts to be detected in VL20 at a temperature as low as 600 °C. From Raman measurements and optical observations, free carbon in low amounts is detectable at 500 °C. In MK and Ceraset, optical observations, EPR analysis and Raman spectroscopy showed that free carbon starts to form at lower temperatures than those detected by means of MAS NMR. Therefore, the free carbon should be present in the VL20 heat-treated at 500 °C.
The solid state MAS NMR spectra are also useful to observe the development of the crosslinking. VL20 heat-treated at 200 °C is only partially crosslinked (the sample is gel like), as also attested by ^{1}H and ^{13}C NMR spectra. At 300 °C the crosslinking is completed. The crosslinking extent is visible in ^{29}Si NMR, where the coordination of silicon can be monitored. At 200 °C only [SiHCN$_2$] and [SiC_2N_2]

units are detected, at 300 and 400 °C [SiC_2N_2] and [SiCN$_3$] (actually at 400 °C only [SiC_2N_2] is visible), and from 500 °C [SiC_2N_2], [SiCN$_3$] and [SiN$_4$].

6.2.2.8. Discussion

The VL20 is a polysilazane similar to Ceraset, and the considerations done on the photoluminescence properties of Ceraset are also valid for VL20. The only difference in their structures is the presence of urea groups in Ceraset. Although the correlation between accelerated crosslinking and presence of urea groups was not demonstrated, the Ceraset shows a faster crosslinking than VL20. The difference in crosslinking rate influences the fluorescence emission spectra. After treatment at 200 °C Ceraset is solid, although not completely crosslinked (see solid state NMR in Figure 51), and its emission is stronger than that of VL20 heat-treated at 200 °C, which is in a gel state and only minimally crosslinked (see solid state NMR in Figure 71). At 300 °C the structure of Ceraset is highly crosslinked and a low amount of mass was lost (5%) (see TGA in Figure 47). The emission from this sample presents the same maximum as the sample heat-treated at 200 °C, with higher intensity. As explained in the section on Ceraset, the presence of the same fluorescence peak at 200 and 300 °C could correspond to limited changes in the structure and the increase in emission intensity could be due to the continued crosslinking through vinyl groups (hydrosilylation). In the case of VL20 the crosslinking is slower. At 300 °C the mass loss of VL20 is over 20 % (see TGA in Figure 68), mainly due to elimination of oligomers, which can freely evaporate due to the limited crosslinking. The emission intensity is very low in the VL20 starting polymer and after treatment at 200 °C it is only slightly stronger (Figure 57). At 300 °C the crosslinking is completed and a stronger emission is observed in the photoluminescence spectrum. Starting from 300 °C, similar photoluminescence behaviors are observed for Ceraset and VL20 and red-shifted emission spectra are observable in both cases. The crosslinking of the VL20 up to 400-500 °C and the formation of sp^2 free carbon from 500 °C are supposed to contribute to the development of the photoluminescence properties. For the discussion on sp^2 free carbon as source of photoluminescence properties the reader is referred to the discussion on MK (paragraph 6.1.1.9.).

The presence of carbon radicals in not excluded, although their role has not been clarified.

6.2.3. KiON S

KiON S is a commercial polysilazane from the company KiON Corp., which is available dissolved in petroleum distillates. The solution is liquid, clear and slightly yellow. Only 10 % of the solution is composed of cyclosilazane with hydrogen and methyl substituents. The remaining 90% of the solution is composed of terpene thinner (hyrocarbons) (22.5-27.0 %), mineral spirits (aromatic naphtha) (18.0-22.5 %) and aromatic petroleum distillates (45.0-49.5 %) [KiON]. The chemical formula declared by the company is illustrated in Figure 72.

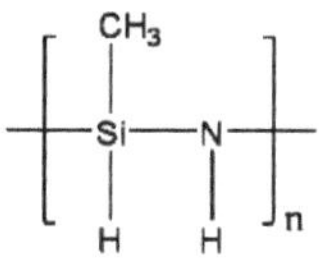

Figure 72: Molecular structural unit of KiON S as provided by KiON Corp.

Crosslinkable groups are provided by the Si-H and N-H bonds.
After removal of the solution, the dried polymer was heat-treated at 200, 300, 400 and 500 °C for 2 h under Ar flow, with heating rate 50 °C/h and free cooling. After drying, KiON S is a hard solid and it is difficult to get a powder out of it. During heat-treatment at different temperatures, the polymer does not melt. The absence of melting hinders the mouldability of the polymer. The creation of porosity inside the KiON S particles during heat-treatment allows the grinding of samples annealed at higher temperatures.

6.2.3.1. Photoluminescence measurements

6.2.3.1.1. Fluorescence measurements

Both the starting and the heat-treated polymers show photoluminescence emission. Figure 73 displays the excitation and emission spectra of KiON S annealed at different temperatures. The excitation and emission wavelengths used for the measurements correspond to the maximum intensity peaks (Table 19).

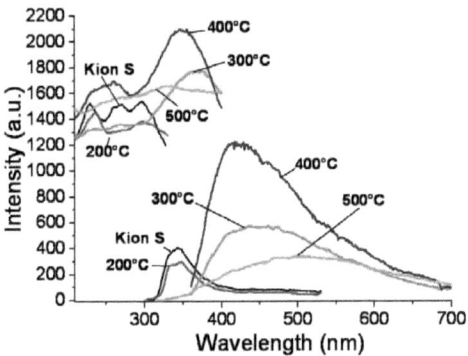

Figure 73: Fluorescence emission (bottom right) and excitation (top left) spectra of KiON S samples obtained with excitation and emission wavelengths according to Table 19.

Table 19: Excitation and emission wavelengths used for the polysilazane KiON S samples.

Sample	Excitation wavelength (nm)	Emission wavelength (nm)
KiON S	230	340
KiON S 200 °C	238	349
KiON S 300 °C	346	435
KiON S 400 °C	346	413
KiON S 500 °C	262	504

Figure 74 represents the fluorescence emission spectra of the polymer KiON S heat-treated at different temperatures, obtained with excitation wavelength of 250 nm (a) and 360 nm (b), respectively.

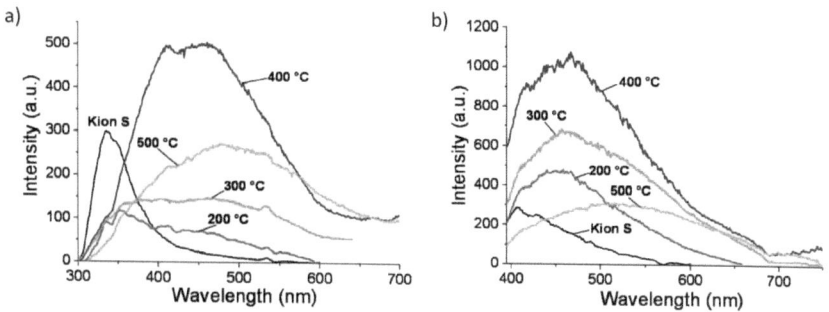

Figure 74: Fluorescence emission spectra of KiON S annealed at different temperature obtained with excitation wavelength of 250 nm (a) and 360 nm (b).

The KiON S samples annealed at different temperatures are shown in Figure 75 under white (a) and UV (360-400 nm) (b) light.

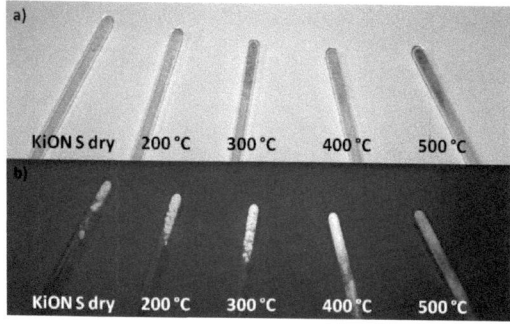

Figure 75: Photographs representing KiON S samples under white (a) and UV (b) light. For colored pictures the reader can refer to http://tuprints.ulb.tu-darmstadt.de/2085/.

The dry KiON S has a yellow coloration under white light and shows maximum fluorescence emission in the UV/blue range. Under excitation at 250 nm it emits in the UV range, while after excitation at 360 nm it shows a blue-greenish emission. After treatment at 200 °C, the maximum fluorescence emission is similar to that of the dry KiON S, while at 360 nm excitation the emission in the blue-green range results more intense than that of the dry KiON S. At 250 nm excitation a weaker emission in the blue/green is observed. After heat-treatment at 300 °C, a broad emission in the visible range with maximum peak at 435 nm is detected as the maximum emission, using excitation wavelength 346 nm. Also with excitation at 250 nm and 360 nm a broad emission in the visible is obtained, and a white-blue/green emission can be observed under UV light. After annealing at 400 °C, the highest emission intensity is detected. The maximum excitation peak is at 346 nm, as after annealing at 300 °C, and the maximum emission peak is at 413 nm. Also in this case, a white-blue/green emission is observed under UV light. Also for excitations at 250 and 360 nm, the highest emission intensity is obtained, with broad spectra in the visible range. After heat-treatment at 500 °C, the sample turns brown and the fluorescence intensity decreases, as previously observed for MK at 700 °C and for Ceraset and VL20 at 600 °C, due to the formation of free carbon. The maximum emission peak is red-shifted to 504 nm.

The spectral emissions obtained with 360 nm excitation of the KiON S samples heat-treated at 200-500 °C are represented on the CIE chromaticity diagram 1931 (Figure 76).

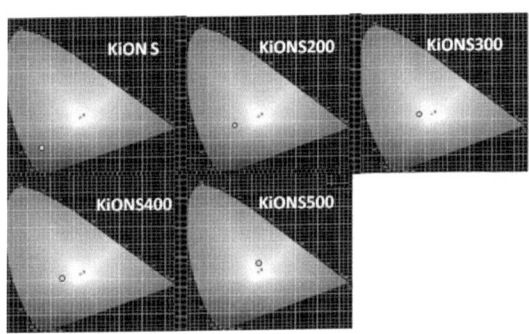

Figure 76: CIE chromaticity diagrams 1931 of the KiON S samples annealed at 200-600 °C. For colored pictures the reader can refer to http://tuprints.ulb.tu-darmstadt.de/2085/.

The color coordinates are perfectly in agreement with the colors of the samples under UV light (Figure 75).

6.2.3.1.2. Stability of the fluorescence during storage

In order to monitor the change in fluorescence properties of the KiON S samples following storage in air, the fluorescence emission spectrum obtained with 250 nm excitation wavelength was detected in the first week after the heat-treatment and about one year later (Figure 77). For the KiON S the time of the storage in air was considered after drying.

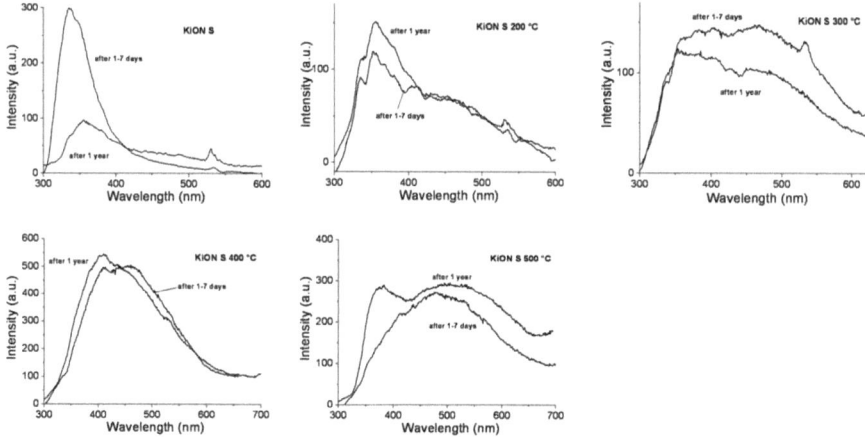

Figure 77: Fluorescence spectra of the KiON S samples obtained just after heat-treatment and about 1 year later (at 250 nm excitation).

The fluorescence of the dried KiON S is significantly decreased after 1 year in air. After heat-treatment at 200, 300 and 400 °C only slight changes in the emission are detectable. KiON S heat-treated at 500 °C after 1 year in air shows the presence of a new peak in the UV range. The hydrogen substituents on the silicon are very sensitive to air. Since the samples were stored in air from the beginning, they probably reacted immediately with air and contained some oxygen also on their first measurement after 1-7 days. While the samples treated at 200, 300 and 400 °C did not change after the first reaction with air, the dried polymer was negatively affected by further storage in air.

6.2.3.1.3. Quantum efficiency

The quantum efficiency was not measured on any KiON S samples, thus it was estimated by comparing the integral of their emission spectra (in energy scale) with the integral of the emission spectra of samples of known quantum efficiency with emission in the same range (Table 20).

Table 20: Quantum efficiencies of the KiON S samples.

Sample	Quantum efficiency (%)
KiON S	5.4*
KiON S 200 °C	4.0*
KiON S 300 °C	7.1
KiON S 400 °C	18.5
KiON S 500 °C	3.3

The samples designed with * emit in the UV range, for which no measurement was successful, thus the value was calculated referring to a spectrum with emission maximum at 394 nm. Since the emission range of KiON S 400 °C was comprised between those of Ceraset 500 °C and VL20 400 °C, its quantum efficiency was calculated considering both samples.

6.2.3.2. Absorption measurements

Additional information about the electronic transitions occurring in the KiON S samples and their variation after temperature treatment can be obtained with absorption measurements.

6.2.3.2.1. UV-Vis-NIR spectroscopy

As previously attested for other heat-treated silicon-based polymers, the solubility of KiON S decreases as the annealing temperature increases. Reasonable UV-Vis spectra were obtained only for the dry KiON S and KiON S annealed at 200 °C, in THF solution (Figure 78).

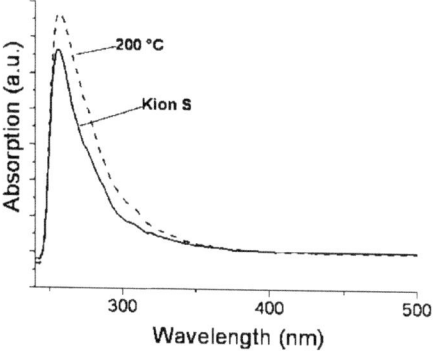

Figure 78: UV-Vis absorption spectra of dry KiON S (solid line) and KiON S annealed at 200 °C (dashed line).

The absorption edges of KiON S and KiON S 200 °C are both around 400 nm (3.10 eV). The spectrum relative to KiON S 200 °C is slightly red-shifted in relationship to that of KiON S.

6.2.3.2.2. Remission measurements

The absorption/scattering curves (Kubelka-Munk function) of the KiON S samples obtained with remission measurements are shown in Figure 79.

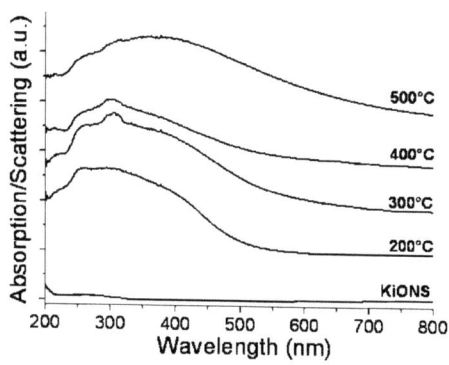

Figure 79: Absorption/scattering spectra of KiON S samples obtained with remission measurements.

A correspondence between the absorption spectra obtained by means of remission measurements and fluorescence spectra can be observed. Both the emission maxima of the fluorescence spectra and absorption edges of absorption/scattering spectra are red-shifted as the heat-treatment temperature increases from 200 to 300 °C, no red-shift is observed from 300 to 400 °C, and a slight red-shift is observed from 400 to 500 °C. The spectrum of the dried KiON S is probably not reliable, because it should be similar to that of KiON S 200 °C, as shown by UV-Vis measurements. Moreover, from UV-Vis measurements, KiON S 200 °C showed an absorption edge of 400 nm, while from remission measurements the edge is at 500 nm (2.48 eV). The absorption edge of KiON S 300 and 400 °C occur at about 550 nm (2.25 eV), for KiON S 500 °C above 700 nm (1.77 eV).

6.2.3.2.3. Reflection measurements

The reflection spectra were obtained using synchronous excitation and detection wavelengths with the Cary Eclipse Varian. The absorption/scattering spectra obtained from the reflection data relative to the KiON S samples (using the Kubelka-Munk function) are presented in Figure 80. The spectra were smoothed for representative reasons. The sample KiON S 400 °C was also normalized, as its intensity was significantly higher than those of the remaining samples.

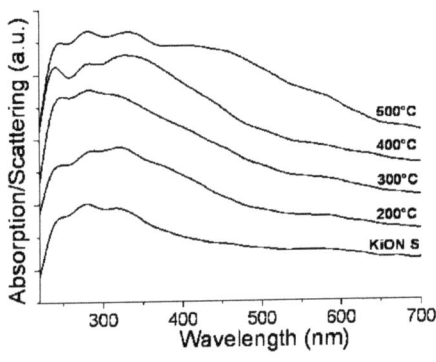

Figure 80: Absorption/scattering spectra relative to the KiON S samples obtained with reflection measurements.

The absorption edges are red-shifted as the heat-treatment temperature increases. KiON S 300 and 400 °C show the same absorption edge, as from remission measurements and in accordance to the photoluminescence spectra.

6.2.3.3. FT-IR spectroscopy

The KiON S and the heat-treated KiON S samples were investigated by means of FT-IR structural analysis (Figure 81), using the ATR device. In Table 21, the vibration bands and respective bonds are listed.

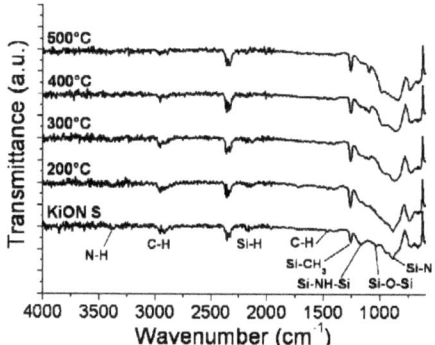

Figure 81: FT-IR spectra of the KiON S samples.

Table 21: Vibration bands and respective bonds detected by means of FT-IR analysis of the KiON S samples.

Bond assignment	Vibration band (cm^{-1})
$v(N-H)$	3379
$v_{as}(-CH_3)$	2956
$v_{as}(CH_2)$	2917
$v(Si-H)$	2169
$\delta_{as}(C-H)$	1472
$\delta_s(Si-CH_3)$	1258

ν(Si-NH-Si)	1160
ν(Si-O)	1036
ν_{as}(Si-N)	888
δ_s(Si-C)	725

Although the chemical formula proposed by the supplier contains many Si-H bonds, from FT-IR analysis Si-H bonds are only slightly detectable. This is due to the storage in air of the samples, with subsequent hydrolysis of the Si-H groups and formation of Si-O bonds, which are detectable in all samples at around 1036 cm^{-1}. The N-H and C-H bonds are only slightly detectable in the precursor and disappear as the annealing temperature increases. The FT-IR measurements do not highlight any significant changes in the structure during heat-treatment, except the reduction of Si-NH-Si groups and the increase in intensity of the Si-O bands.

6.2.3.4. Raman spectroscopy

The heat-treated KiON S samples were investigated with confocal micro-Raman spectrometer, using an excitation laser of 633 nm, but only fluorescence interference was detected for all samples. Thus, the samples were also analyzed with the IR/Raman spectrometer Bruker IFS 55 - FRA 106, with laser wavelength 1064 nm. The spectra are reported in Figure 82 and in Table 22 the vibration bands and respective bonds are listed.

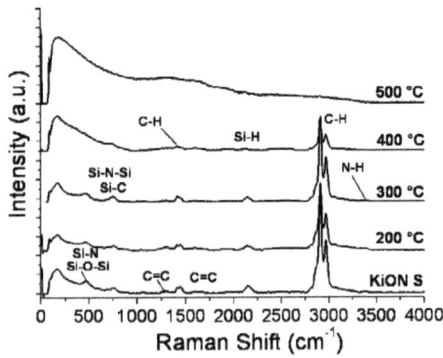

Figure 82: Raman spectra of the KiON S samples using 1064 nm as laser wavelength.

Table 22: Vibration bands and respective bonds detected by means of FT-IR analysis of the KiON S samples.

Bond assignment	Raman shift (cm^{-1})
$v(N-H)$	3390
$v_{as}(-CH_3)$	2966
$v_{as}(CH_2)$	2906
$v(Si-H)$	2146
$v(C=C)$	1595-1664
$\delta(C-H)$	1412
$v(Si-N-Si)+v_s(Si-C)$	756
$v(Si-O-Si)+v(Si-N)$	467

No luminescence interference is present in the measurements; however, KiON S 500 °C shows interference due to the heating of carbon clusters, as previously observed in MK 700 °C.

The Si-H and N-H bands decrease in intensity as the treatment temperature increases. Si-O-Si and Si-N bonds are detectable. Many weak bands are present, in the precursor and after heat-treatment up to 400 °C, in the range from 1300 cm^{-1} to 1590 cm^{-1}, denoting C=C bonds. Since the KiON S structure does not contain C=C bonds, they are assigned to residues from the aromatic solvent.

6.2.3.5. TGA/DTG /MS

The monitoring of the decomposition behavior of the dried polymer KiON S was performed by means of TG/DTG/MS (Figure 83 and Figure 84).

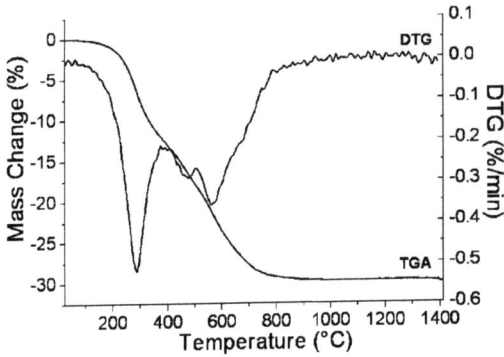

Figure 83: TGA and DTG of KiON S after drying.

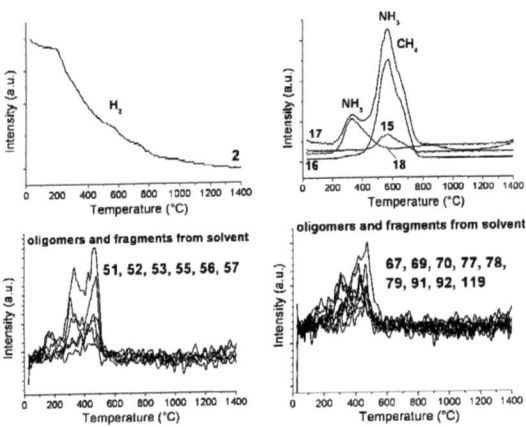

Figure 84: Selected TG/MS analysis of the gaseous byproducts of the thermal decomposition of KiON S after drying. The numbers in the graphs refer to m/z.

The TGA and DTG graphs of the dried polysilazane KiON S display three main mass losses. As MS analysis shows, the first decomposition step, up to 400 °C, corresponds to the elimination of nitrogen-derived species, NH_4^+ and NH_3 due to the cleavage of Si-N bonds. Moreover, H_2 and oligomers are also eliminated. The second decomposition step (400-520 °C) corresponds to the mineralization with evolution of a small amount of gases (H_2, NH_3 and CH_4). Finally, the third decomposition step, corresponding to the ceramization process, is due to the formation of volatile gases (H_2, NH_3 and CH_4). Moreover, up to 500 °C, fragments from the solvent not extracted from the polymer (z/e=51-119) evaporate. Both oligomers and fragments from the solvent could be responsible for the detected species with higher masses. Since these species outnumber those detected in Ceraset and VL20, they should be in great part related to the solvent evolution. As previously mentioned, our MS analysis does not consider evolution of species with even higher masses, for example naphthalene, which constitute the solvent.

Besides Raman spectra, MS offers a further proof of the presence of solvent in the KiON S samples heat-treated up to 500 °C.

6.2.3.6. XRD measurements

The XRD analysis of the KiON S heat-treated at different temperatures is displayed in Figure 85.

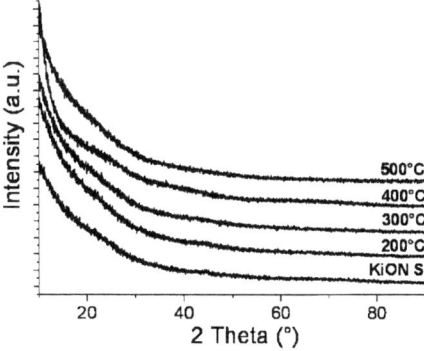

Figure 85: X-ray diffraction patterns of KiON S samples heat-treated at different temperatures.

The absence of crystalline peaks evidences the amorphous character of the material in the studied temperature range.

6.2.3.7. Liquid state NMR Spectroscopy (^{1}H, ^{13}C and ^{29}Si)

In order to determine the species present on the polymer after drying, NMR analysis was carried out on the dried KiON S, measured in C_6D_6 as deuterated solvent. The graphs representing ^{1}H, ^{13}C and ^{29}Si DEPT (Distortionless Enhancement by Polarization Transfer) are displayed in Figure 86.

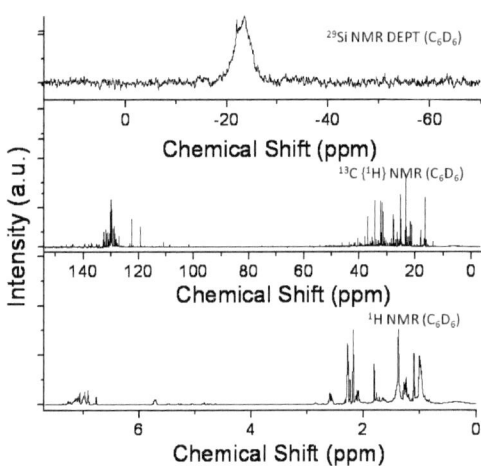

Figure 86: ^{1}H, ^{13}C and ^{29}Si DEPT NMR of KiON S after drying.

From ^{13}C NMR analysis it is clear that the dried KiON S still contains solvent traces, as attested by the peaks due to aromatic compounds (150-100 ppm). The ^{29}Si NMR DEPT shows a single peak at -23.5 ppm, indicating the presence of a single environment for the silicon, i.e. that illustrated by the chemical structure in Figure 72.

Due to the presence of solvent in KiON S, this polymer is not relevant for the analysis on silicon-based luminescent polymers. Therefore, solid state NMR analysis was not performed, because of the time consuming character of the measurement.

6.2.3.8. Discussion

Although the structure of KiON S is quite similar to the previously analyzed polymers, the polymer shows different properties, mostly related to its availability in solution. Compared to the previously analyzed polymers, KiON S displays photoluminescence emission in the visible range at lower treatment temperatures and even in the untreated polymer (Figure 73). This is an advantage for LED applications, where visible emission is desired after annealing temperatures inferior or equal to 300 °C. Moreover, the calculated quantum efficiency of the sample annealed at 400 °C is the highest found until this point of the thesis for visible emissions (Table 20).

Nevertheless, this polymer was not further analyzed for two main reasons. First of all, since the polymer is available only in solution, it is unpractical for the purpose that we are studying. It must be dried, resulting in solid particles which are not meltable and moldable. In the case of Ceraset and VL20, the liquid polymers could be poured in the mould and heated until they crosslink and form a shaped solid. In the case of MK, the solid polymer melts upon heating in the mold and subsequently crosslinks. Direct molding is for KiON S impossible. Its application could be solution molding, i.e. molding of the polymer in solution and subsequent slow evaporation of the solvent. The dry KiON S, reduced to 1/10 of its initial volume could be shaped and afterwards heat-treated to the suitable temperature for the required emission properties.

The second reason why KiON S was not further analyzed is the contamination of the polymer by the solvent, as proven by liquid state NMR, Raman and MS analyses, which affects the photoluminescence properties and is probably the origin of them (Figures 86, 82 and 84). The aromatic compounds present in the solvent, for example naphthalene, are luminescent and stable up to high temperatures [Bolbit2000, Eremenko1969]. Although the solvent was evaporated from the polymer and a heat-treatment was performed, some residues remain inside after the heating, as proven by MS analysis and Raman spectra, and could be responsible for the photoluminescence properties even at relatively high annealing temperatures. Therefore, the analysis of the luminescence mechanisms of the KiON S is not straightforward and it is not possible to exclude the activity of the aromatic solvent as source of luminescence. For this reason, in the study of the source of the photoluminescence properties in silicon-based polymers, KiON S was not considered.

Nevertheless, the interesting fluorescence properties of this polymer were presented here, which could derive both from the effect of the solvent or from the polymer itself and its heat-treatment. The KiON S possesses potential for solution molding applications, which are slightly different from those of the remaining silicon-based polymers.

6.3. Polysilylcarbodiimides

6.3.1. Phenyl-Containing Polysilylcarbodiimides

Four polyphenylsilylcarbodiimide derivatives, namely $-[PhRSi-NCN]_n-$ (S1-S4), were synthesized by the reaction of phenyl-containing dichlorosilanes with bistrimethylsilylcarbodiimide, in the presence of pyridine as catalyst, as pointed out in the experimental part. The first substituent on the silicon is phenyl in all polymers, while the second substituent is varied between hydrogen, methyl, vinyl and phenyl [Mera2009b, Morcos2008b].

Phenyl containing polymers have been shown to provide luminescence properties [Suzuki1997, Salom1987, Dias2000].

The four polysilylcarbodiimide derivatives were heat-treated in argon at different temperatures, from 200 up to 500 °C for 2 h under Ar flow, with a heating rate of 50 °C/h and free cooling. After low temperature treatments the polymers showed distinctive luminescence properties, which could be interesting for applications such as LEDs. Furthermore, the polymers S1-S4 are soluble in all common solvents and therefore easy to shape either as synthesized or after heat-treatment up to 400 °C.

The molecular structures of the four phenyl-containing polysilylcarbodiimides investigated in this study are illustrated in Figure 87. The polysilylcarbodiimides with the second substituents being phenyl, methyl, hydrogen and vinyl are denoted as S1 (polydiphenylsilylcarbodiimide), S2 (polyphenylmethylsilylcarbodiimide), S3 (polyphenylsilylcarbodiimide) and S4 (polyphenylvinylsilylcarbodiimide), respectively. Although the air sensitivity decreases if bulky aromatic substituents are present at the silicon, they are still slightly air and moisture sensitive. Therefore, the polymers and the heat-treated samples were consequently stored in argon.

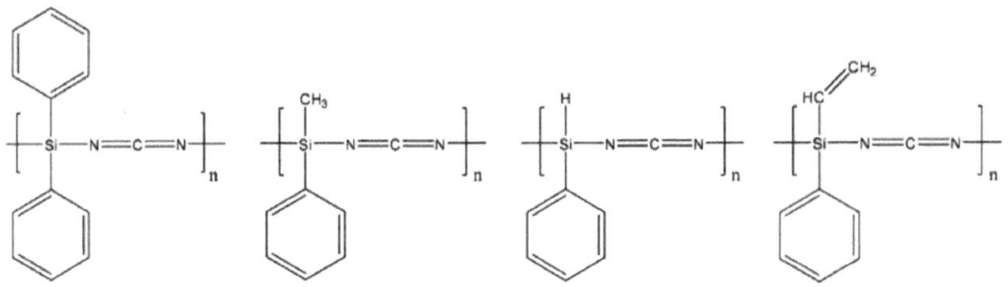

6. Results and discussion

Figure 87: Molecular structural units of the four polysilylcarbodiimides S1, S2, S3 and S4.

Although not displayed in the graphs, end groups are present in the polymers: $-SiMe_3$ in all polymers, $-SiPh_2Cl$ in S1, -SiPhMeCl in S2, -SiPhHCl in S3 and -SiPhViCl in S3, as published by Mera et al. [Mera2009a].

These polymers are not commercially available and are not extensively used and known as the MK, Ceraset or VL20. Nevertheless, they have been largely studied in our group, and the availability of several investigations on low temperature treated samples, besides the capability of controlling the synthesis, renders them suitable candidates for our purpose. While in commercial polymers the chemical structure and possible contaminating compounds cannot be controlled, the synthesis of these polymers is controlled. Moreover, the phenyl groups on the silicon element of the chain are luminescent substituents, which offer a specific emission that can be monitored after heat-treatment.

6.3.1.1. Photoluminescence measurements

6.3.1.1.1. Fluorescence measurements

6.3.1.1.1.1. Photoluminescence properties of the polymers

In Figure 88, the normalized excitation and emission spectra of the four polymers are shown (normalization at 272 nm). The spectra were obtained with the excitation and emission wavelengths relative to the maximum emission and excitation intensities (Table 23).

6. Results and discussion

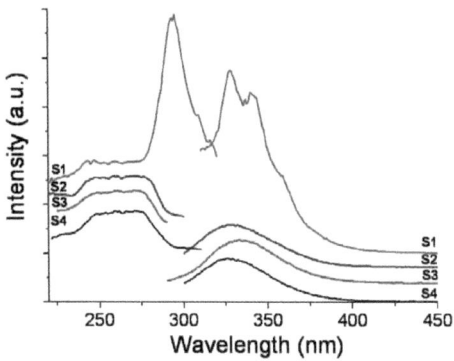

Figure 88: Fluorescence emission (bottom right) and excitation (top left) spectra of the polysilylcarbodiimides S1-S4. The spectra are vertically shifted for clarity.

Table 23: Excitation and emission wavelengths used for the polysilylcarbodiimides. They are the maximum excitation and emission peaks.

Sample	Excitation wavelength (nm)	Emission wavelength (nm)
S1	290	343
S2	272	328
S3	272	335
S4	270	329

The polymers with hydrogen, methyl and vinyl as the second substituent show similar excitation and emission spectra. They all present three excitation peaks at 245 nm, 259 nm and 275 nm. Their emission spectra are composed by a single peak centered at 329 nm, 334 nm and 328 nm for S2, S3 and S4 respectively. Thus, S2, S3 and S4 are characterized by a single fluorescence emitting species that emits at around 330 nm and by three excitation maxima. The exact emission wavelength is slightly shifted in the different polymers. The excitation spectra of the four polymers were normalized at 272 nm, so that the contribution of the three common excitation peaks results the same for all polymers. The same normalization was applied also to the emission spectra.

Besides the mentioned peaks, the polymer S1 shows a new and more intense excitation peak at 294 nm. The emission spectrum of S1 is composed of two contributions: the first is the same emission peak present in the other polymers at 328 nm, which is excited by the three common excitation peaks; the second is an emission with vibrational structure centered in the range between 328 and 342 nm, excited by the excitation band at 294 nm.

A comparison with the literature shows that the emission peak located at 328 nm corresponds to the excimer transitions of phenyl groups. Accordingly, polymethylphenylsiloxane excimer emission was detected at 328 nm [Suzuki1997], 325 nm [Salom1987] and 320 nm [Dias2000]. No reports were found on the luminescence properties of polyphenylsiloxane and polyphenylvinylsiloxane. Also polystyrene with phenyl pendant groups shows photoluminescence emission at around 328 nm [Salom1987].

To prove the inactivity of the carbodiimide group as part of the backbone of the Si-polymer in the photoluminescence, the phenyl-free polydimethylsilylcarbodiimide was also characterized and did not show photoluminescence properties at room temperature. Therefore, the common emission peak present in all polysilylcarbodiimides is assigned to the excimer transition of the phenyl groups. The N=C=N backbone units of the polymers do not contribute to the measured luminescence emission and do not interact with the luminescence mechanism of phenyl groups. The emission wavelength is blue-shifted for larger second substituents on the silicon atom, from hydrogen to methyl to vinyl groups, due to steric reasons.

In polymethylphenylsiloxane, excimers are formed by neighboring phenyl rings, whose distance is high enough that the interaction between contiguous phenyl rings in their ground states is attractive. This attraction between neighboring rings favors the excimer emission [Salom1987]. In the case of the polysilylcarbodiimides, the neighboring phenyl groups are separated by a rigid N=C=N unit. No phenyl monomer emission was detected. The large distance between contiguous chromophores could favor the excimer emission. Nevertheless, the rigid N=C=N unit could also hinder the formation of excimers from neighboring phenyl groups, because of a deficient π-overlapping, and thus the excimers could be formed by phenyl rings which are far away in the same chain or come from another chain, thanks to the polymer flexibility. Thus, since the polymer is not diluted, the transition could be intramolecular as well as intermolecular. In S2, S3 and S4 the only observable transition is the excimer emission of the phenyl groups.

In polydiphenylsilylcarbodiimide, the second substituent at the silicon is another phenyl group, influencing the fluorescence emission. In addition to the peak relative to the phenyl excimer emission and its three excitation peaks, a new emission with vibrational structure in the range 328-348 nm and a new excitation peak at 294 nm are detected. The new transitions should be correlated to the simultaneous presence of two phenyl groups at the same silicon atom, and their interaction, otherwise the excimer emission at 328 nm would be emphasized, but no new peaks would be detected.

In Figure 89, the emission spectra of S1 obtained with different excitation wavelengths (250, 290, 300 and 360 nm) are shown.

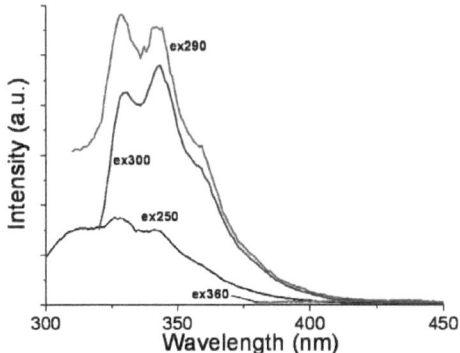

Figure 89: Fluorescence emission spectra of the polydiphenylsilylcarbodiimide S1 obtained with different excitation wavelengths (250, 290, 300 and 360 nm).

Considering Figure 89 and the simulations performed in Matlab, the emission spectrum results composed of two emitters: a Gaussian peak, corresponding to the phenyl excimer emission, and an emission with vibrational structure. With 250 nm as the excitation wavelength, the contribution of the excimer emission prevails, but the contribution of a different center is appreciable. With exciting wavelengths 290 and 300 nm, which correspond to the excitation maximum (294 nm), the second center with vibrational structure is dominant.

The peak with the vibrational structure, present only in S1, is assigned to the monomer emission of the delocalized unit Ph-SiX$_2$-Ph (X=NCN). The p orbitals related to the two phenyl units interact as a π-conjugated system through the silicon atom [Suzuki1998, Rathore2009].

In order to observe the emission of the single unit Ph-SiX$_2$-Ph, diphenyldimethylsilane was used as model compound in hexane solution (X=methyl). In Figure 90, the comparison of the normalized emission spectra of S1 and diphenyldimethylsilane is shown at two different excitation wavelengths, 250 nm (a) and 290 nm (b).

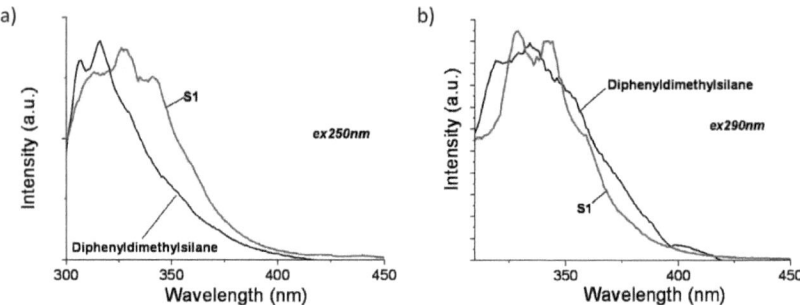

Figure 90: Comparison between the normalized fluorescence emissions of S1 and diphenyldimethylsilane at 250 nm (a) and 290 nm (b) as excitation wavelengths.

In Figure 91, the emission spectra of diphenyldimethylsilane at different excitation wavelengths are shown.

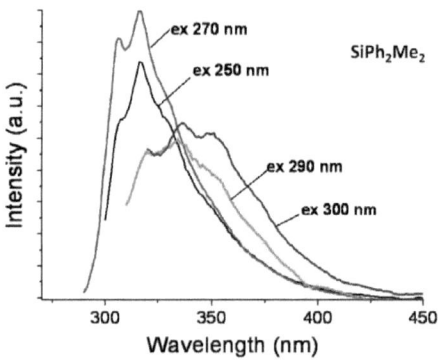

Figure 91: Fluorescence emission spectra of diphenyldimethylsilane at different excitation wavelengths.

As expected, the emission with vibrational structure present in S1 was detected also in diphenyldimethylsilane, but blue-shifted by 0.09 eV. In the case of diphenyldimethylsilane, the Ph-SiX$_2$-Ph monomer emission is highest for an excitation at 250-270 nm, while in S1 it is found at 294 nm. The slightly different absorption and emission energies are related to the degree of interaction between the two phenyl groups, which in turn depends on the nature of the X-group of the Ph-SiX$_2$-Ph unit and on the geometry of the molecules, i.e. the angle between the phenyl planes. In the polymer, more hindrance between substituents is present and thus the angle between the phenyl groups is maximized (180°). However, in the monomer, the angle results below 180° because of the absence of hindrance. As the wavelength emission depends on the conjugation, a red-shift is expected in the case of the polymer since the planar geometry is more favorable for the conjugation [Kim1997, Kim1998]. Diphenylmethane, as a model compound with methylene as the interconnecting unit between two phenyl groups, also shows a vibrational structure due to the presence of two phenyl groups, although without the interposition of a silicon atom [Cooke2003, Salimgaareeva2003]. Therefore, we can attest that the vibrationally structured peak of S1 is related to the simultaneous presence of two phenyl rings attached to the silicon atom, i.e. it coincides with the emission from the unit Ph-SiX_2-Ph. Only polymer S1 possesses this chromophore; S2, S3 and S4 do not, thus they do not show this vibrationally structured emission. The Ph-SiX_2-Ph chromophore is also present in polydiphenylsiloxane, where the same vibrational structure is detected in the emission spectra. In this case, the emission range is red-shifted compared to that of our polydiphenylsilylcarbodiimide, due to the different backbone unit X of the Ph-SiX_2-Ph group [Suzuki1997].

6.3.1.1.1.2. Photoluminescence properties of the heat-treated polymers

As for other silicon-based polymers in previous chapters, the four phenyl-containing polysilylcarbodiimides were heat-treated at relatively low thermolysis temperatures. Their photoluminescence behavior was studied after annealing of the polymers between 200 and 500 °C. The heat-treated polymers showed different physical properties. Polymer S1 behaves rubber-like after temperature treatments up to 400 °C. Polymer S2 remains liquid after heat-treatment up to 300 °C and exhibits a rubber-like texture at 400 ° C. Polymer S3 forms a powder at room temperature and remains in powder form also after heat-treatment up to 500 °C. Polymer S4 is rubber-like after heating at 200 °C and starts to become glass-like via heat-treatment at 300 °C. All polymers turn black glassy materials after treatment at 500 °C.

In Figure 92, the luminescence properties of the four thermally treated polymers, S1 (a), S2 (b), S3 (c) and S4 (d), are shown. The spectra were obtained with the excitation and emission wavelengths corresponding to the maximum emission and excitation intensities (Table 24).

6. Results and discussion

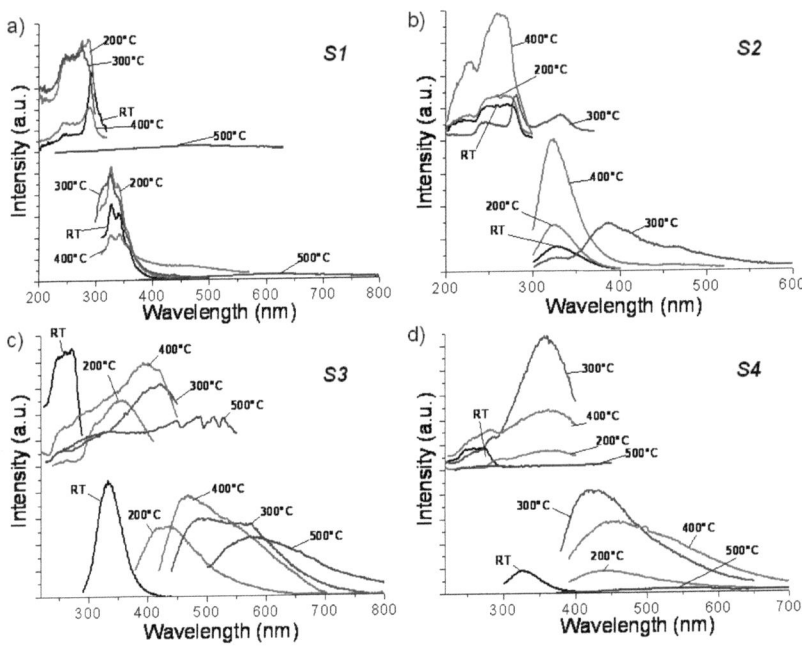

Figure 92: Fluorescence emission (bottom right) and excitation (top left) spectra of S1 (a), S2 (b), S3 (c) and S4 (d) heat-treated at different temperatures with emission and excitation wavelengths as follows from Table 24. The intensities are not comparable.

Table 24: Excitation and emission wavelengths used for the polysilylcarbodiimides samples, corresponding to the maximum excitation and emission peaks.

	S1		S2		S3		S4	
	Exc.	Em.	Exc.	Em.	Exc.	Em.	Exc.	Em.
Room T	290	343	272	328	272	335	270	329
200 °C	288	326	270	327	360	438	370	441
300 °C	276	326	281	391	420	495	360	425
400 °C	290	345	265	327	390	472	364	443
500 °C	333	645	-	-	482	574	300	540

Figure 92 shows how it is possible to obtain a variety of emission ranges by simple heat-treatment. It is possible to individuate two general trends: S1 and S2 essentially show the same emission maximum after heat-treatment up to 400 °C, while S3 and S4 display a red-shift of the emission spectra at increasing temperatures, as previously observed in MK, Ceraset and VL20. The analysis of the spectra will be discussed later, in view of the structural analyses. The sample S2 annealed at 500 °C did not exhibit any fluorescence properties.

The spectral emissions obtained with 360 nm excitation of the polysilylcarbodiimides samples heat-treated at 200-500 °C are represented on the CIE chromaticity diagram 1931 (Figure 93 (a) (b) (c) and (d) for S1, S2, S3 and S4, respectively).

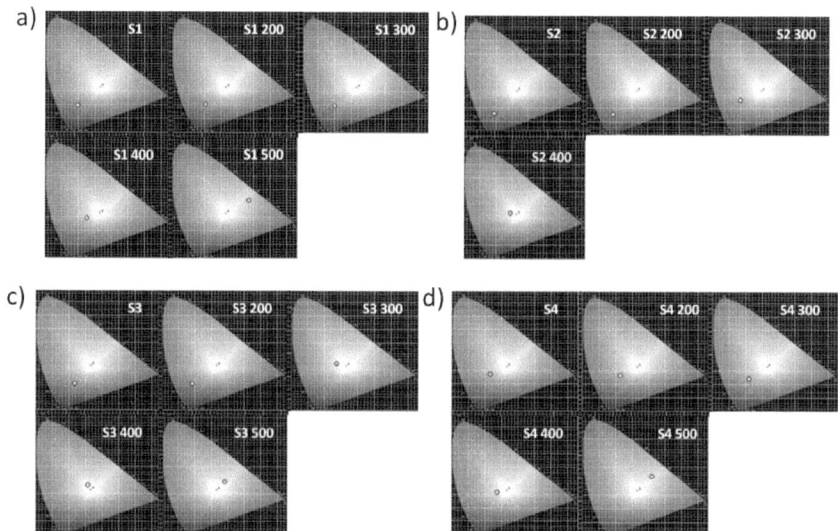

Figure 93: CIE chromaticity diagrams 1931 of the S1 (a), S3 (c) and S4 (d) samples annealed at 200-500 °C and of the S2 (b) samples annealed at 200-400 °C. For colored pictures the reader can refer to http://tuprints.ulb.tu-darmstadt.de/2085/.

The color coordinates are in agreement with the colors of the samples under UV light.

6.3.1.2. Absorption measurements

Additional information about the electronic transitions occurring in the polysilylcarbodiimides samples and their variation after temperature treatment can be obtained with absorption measurements.

6.3.1.2.1. UV-Vis-NIR Spectroscopy

UV-Visible spectroscopy was applied to the polymers and to the heat-treated samples, using tetrahydrofuran (THF) as solvent. In Figure 94, the normalized absorption spectra of the four polysilylcarbodiimides are illustrated.

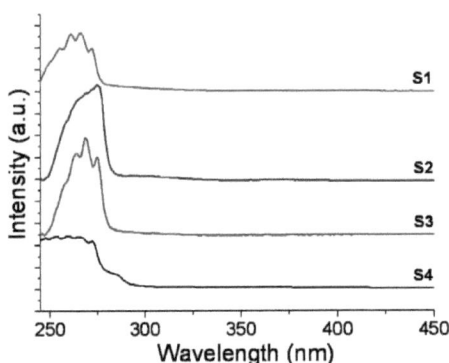

Figure 94: Normalized absorption spectra of the four polysilylcarbodiimides in THF.

All the absorption spectra show vibrational structure and similar shape. The absorption edge of S1 is at about 280 nm (4.43 eV), for S2 at 286 nm (4.34 eV), for S3 is at 285 nm (4.35 eV) and for S4 at 293 nm (4.23 eV). S4 is slightly different from the other spectra, presenting a new shoulder at lower energies. The common absorption peak is due to the phenyl groups. Similar absorption spectra were found for the polydiphenylsiloxane [Bothelho do Rego2001].

As already observed in other heat-treated silicon-based polymers, the solubility of the samples decreases as the treatment temperature increases. The solubility of the different polymers is related to the presence of crosslinking. In Figure 95, the UV-Vis absorption spectra of S1 (a), S2 (b), S3 (c) and S4 (d) heat-treated at different temperatures are presented. Only the spectra of the soluble samples are shown.

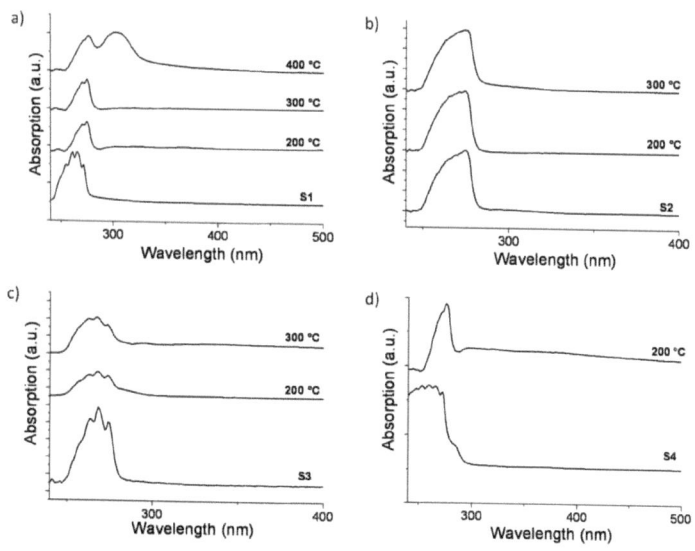

Figure 95: UV-Vis spectra of the S1 (a), S2 (b), S3 (c) and S4 (d) heat-treated at different temperatures. The intensities are not comparable.

The UV-Vis spectra of the heat-treated samples roughly present the same features as the starting polymers. In S1, a new red-shifted peak at 303 nm is detectable at 400 °C, and it could be attributed to the absorption from aromatic agglomerations, which are supposed to develop at that temperature.

Reflection measurements were not performed on polysilylcarbodiimide samples, because of the too low sample amount available.

6.3.1.3. FT-IR spectroscopy

The polysilylcarbodiimides S1 (a), S2 (b), S3 (c) and S4 (d) and the heat-treated samples were investigated with the FT-IR structural analysis using the ATR device (Figure 96). In Table 25 the vibration bands and respective bonds are listed.

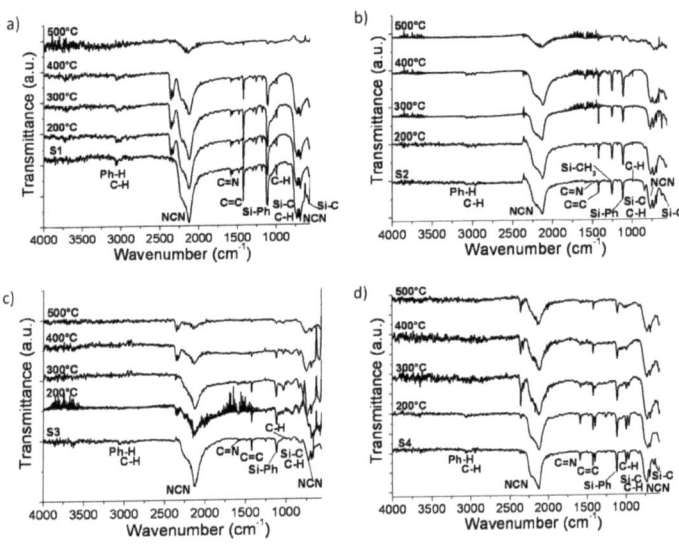

Figure 96: FT-IR spectra of the polysilylcarbodiimides S1 (a), S2 (b), S3(c) and S4 (d) and the heat-treated samples.

Table 25: Vibration bands and respective bonds detected by means of FT-IR analysis of the polysilylcarbodiimides samples.

Bond assignment	Vibration bands (cm^{-1}) for S1/S2/S3/S4
$\nu(C_{arom}-H)$	3071/3068/3058/3069
$\nu(C-H)$	2924/2956/2959/2948
$\nu_{as}(NCN)+\nu(Si-H)$ in S3	2124/2128/2128/2128
$\nu_s(NCN)+\nu(C=C)$	1590/1590/1592/1595
$\nu(C=C)$	1428/1429/1429/1429
$\delta_s(Si-CH_3)$ in S2	1258
$\nu_s(Si-Ph)$	1112/1121/1121/1120
$\delta(C-H)$	1067/1063/1065/1064
$\delta(C_{arom}-H)$	998/999/999/1000
$\delta(Si-C)$	853/847/826/846
$\delta(NCN)$	760/773/773/766
$\gamma(NCN)$	736/736/729/738
$\nu(Si-C)$	692/695/691/697
$\nu(Si(NCN))$	581/573/573/579

The FT-IR bands relative to S1 samples remain almost unvaried up to 400 °C and after heat-treatment at 500 °C the signals are significantly reduced. Accordingly, the polymer does not crosslink up to 400 °C and degrades at around 500 °C, turning black and causing a reduced transmission signal. Also in polymer S2, few changes are detectable up to 400 °C and the intensity of the signals at 500 °C is reduced. Remarkable is the presence of the Si-CH$_3$ peak (1258 cm^{-1}) up to 500 °C. In polymer S3, at 500 °C the signals are significantly reduced, due to the reduced transmission of the black ceramic sample. The Si-H band overlaps the NCN band; therefore it is not possible to monitor the crosslinking of the polymer through Si-H groups by means of FT-IR. In the FT-IR spectra of polymer S4, as the annealing temperature increases, the C=C bonds relative to vinyl groups decrease. Thus, the polymer S4 crosslinks through its vinyl groups. As no other crosslinking possibilities are present, vinyl polymerization is the crosslinking mode.

In all polymers, Si-CH$_3$ bonds are detected, relative to the end groups. The N=C=N bonds are still detectable in all polymers at 500 °C.

6.3.1.4. Raman spectroscopy

The polysilylcarbodiimide samples S1 (a), S2 (b), S3 (c), S4 (d) and the heat-treated samples were investigated by means of IR/Raman spectrometer Bruker IFS 55 - FRA 106, with laser wavelength 1064 nm (Figure 97 and Table 26).

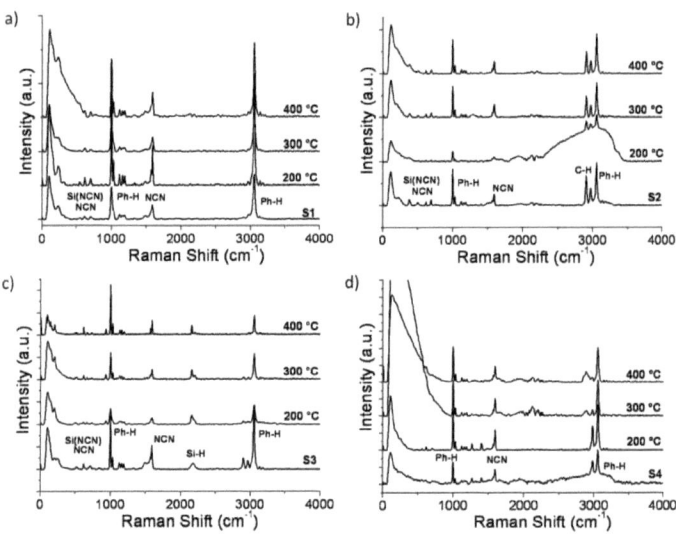

Figure 97: Raman spectra of the polysilylcarbodiimides S1 (a), S2 (b), S3(c) and S4 (d) and heat-treated samples using 1064 nm as laser wavelength.

Table 26: Vibration bands and respective bonds detected by means of Raman analysis of the polysilylcarbodiimides samples.

Bond assignment	Raman shift (cm^{-1})
$\nu(Ph\text{-}H)$	3051
$\nu_s(CH_3)$ in S2	2901/2905
$\nu_s\,(\text{-}CH_2\text{-})$ in S4	2883
$\nu(Si\text{-}H)$ in S3	2136, 2154
$\nu_s(NCN)$	1593/1587
$\delta_{ip}(Ph\text{-}H)$	994/998
$\nu_{as}(Si(NCN)) + \gamma(NCN))$	614

In S1 and S2 no relevant structural changes were detected up to 400 °C by means of Raman measurements, due to the absence of crosslinking. In polymer S3, the band relative to Si-H bonds is significantly reduced after heating up to 400 °C, indicating the occurrence of crosslinking reactions via Si-H groups. Nevertheless, not all Si-H groups were involved in the reaction, as the Si-H peak is still present at 400 °C. In S4, the peaks at 1402 and 1272 cm^{-1}, present up to 200 °C and denoting unsaturated C-H groups and C=C bonds, disappear at higher treatment temperatures, indicating the reaction of vinyl substituents. A new peak appears at 2883 cm^{-1} from 300 °C, denoting -CH$_2$- formed by the crosslinking reactions of vinyl groups.

The samples heat-treated at 500 °C presented interference because of the heating of carbon clusters and are not reported here.

6.3.1.5. TGA/DTG/MS

Simultaneous thermal analysis was carried out on the four polysilylcarbodiimides and the TGA and DTG curves are illustrated in Figure 98. The thermal transformation of the polymers S1-S4 to carbon rich SiCN ceramics at 1100 °C was previously reported [Mera2009b].

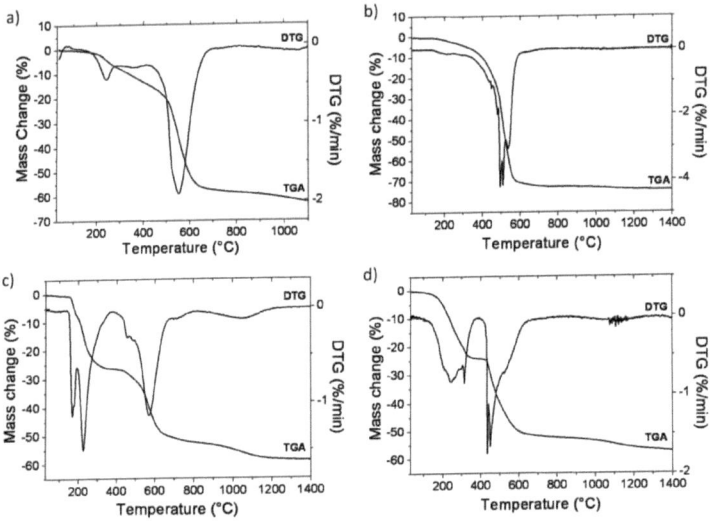

Figure 98: TGA and DTG curves of S1 (a), S2 (b), S3 (c) and S4 (d).

In Figure 99 the TGA curves of the four polysilylcarbodiimides are compared in a single graph.

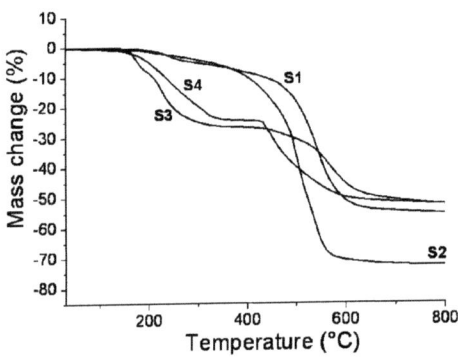

Figure 99: TGA curves of the four polysilylcarbodiimides analyzed.

From both Figure 98 and 99 it is immediately clear that the S1 and S2 present different behavior than S3 and S4. The first two polymers show mass losses lower than 10 % up to 400-500 °C, but after the "transition temperature" the weight loss increases. At 600 °C, their mass loss is between 50 and 70 %. The polymers S3 and S4, on the contrary, start to lose weight at temperatures below 200 °C. At 300 °C, they lost around 25 % of the initial mass. The polymer S3 shows a plateau between 300 and 450 °C, before starting to lose weight again. At 600 °C, its mass loss is around 45 %. Also the polymer S4 shows a plateau, between 350 and 430 °C, before showing a mass loss more abrupt than that of S3 and at 600 °C its weight loss is 50 %.

The two different behaviors, relative to S1 and S2 on one side and to S3 and S4 on the other side, are explained considering that in S1 and S2 no crosslinking takes place up to 500 °C, while the polymers S3 and S4 thermoset through hydrogen and vinyl groups, respectively. Nevertheless, vinyl polymerization does not cause weight loss, and the higher mass loss of S4 compared to S1 and S2 is attributed to a higher end groups content, which leave the material. The mass loss of S3 is instead related to possible dehydrocoupling reactions, besides the end groups elimination [Morcos2008]. The strong mass losses of samples S3 and S4 at low temperatures could indicate the crosslinking of the polymers through end groups, besides the crosslinking through hydrogen and vinyl groups.

In Figure 100, selected MS spectra of gaseous byproducts relative to the thermal transformation of the polymers S1 (a), S2 (b), S3 (c) and S4 (d) are illustrated.

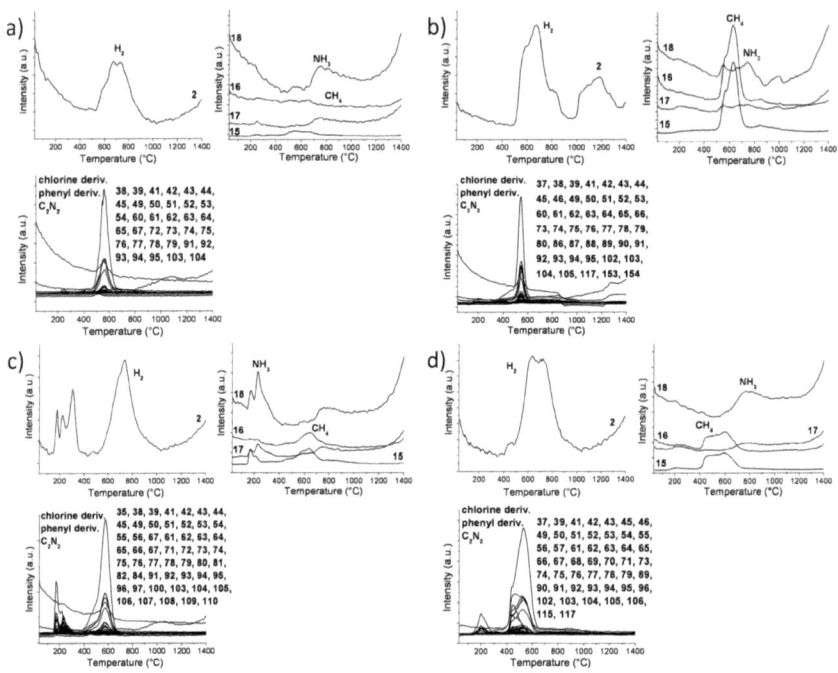

Figure 100: Selected TG/MS analysis of the gaseous byproducts of the thermal decomposition of S1 (a), S2 (b), S3 (c) and S4 (d). The numbers in the graphs refer to m/z.

Except for the elimination of end groups in small amounts (Me$_3$SiCl), S1 and S2 do not lose any compounds up to 400 °C. In accordance with the TGA, the degradation of the polymers start at 500 °C, with loss of H$_2$, phenyl derivatives, chlorine derivatives and C$_2$N$_2$, and the polymers are converted into a ceramic. Furthermore in S2, crosslinking reactions between methyl groups start at about 500 °C, with release of methane.

In the case of S3 and S4, the first mass loss at around 200 °C is associated with the elimination of end groups (Me$_3$SiCl and mainly phenyl derivatives). The additional hydrogen loss in S3 indicates the occurrence of possible dehydrocoupling reactions [Morcos2008]. In the case of S4, the crosslinking reactions between vinyl groups do not cause mass losses. At temperatures higher than 400 °C degradation reactions occur for both polymers.

6.3.1.6. XRD measurements

The X-ray diffractograms of the polysilylcarbodiimide samples heat-treated at different temperatures are displayed in Figure 101.

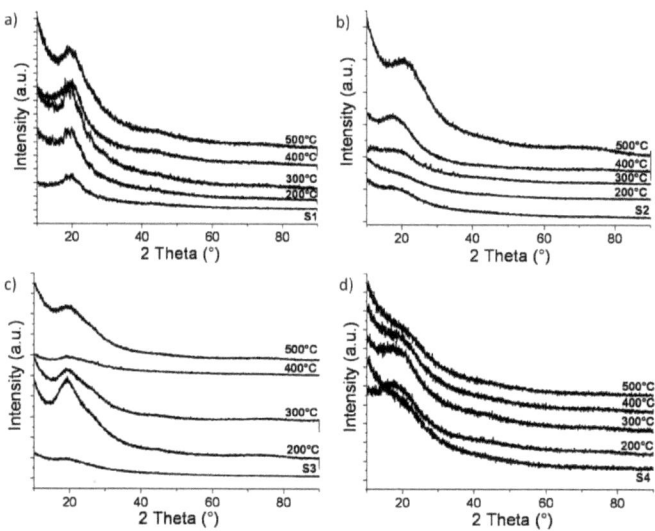

Figure 101: X-ray diffraction patterns of S1 (a), S2 (b), S3 (c) and S4 (d) heat-treated at different temperatures.

Since no crystalline phase is detected with XRD measurements in any of the samples, all the polysilylcarbodiimide polymers remain amorphous after heat-treatment up to 500 °C.

6.3.1.7. Discussion

The photoluminescence features of S1-S4 are analyzed after heat-treatment of the polymers right below the temperature at which conversion into the ceramic material takes place (Figure 92). For polysilylcarbodiimides, this transformation can be formally described by the temperature limit above which elimination of gaseous C_2N_2 is analyzed (T_{S1} = 557 °C, T_{S2} = 545 °C, T_{S3} = 573 °C, T_{S4} = 526 °C), as quantitatively detected by means of TGA, according to the reaction in Figure 102 [Riedel1998].

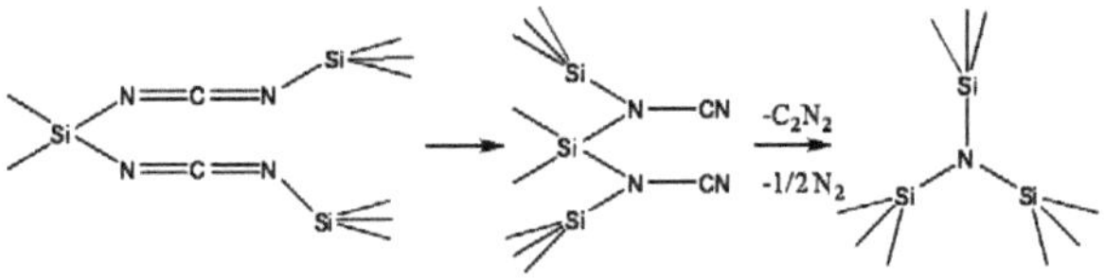

Figure 102: Thermal transformation of a polysilylcarbodiimide during the polymer-to-ceramic conversion.

In the present work, all polymers turned black glassy materials after heat-treatment at 500 °C. Therefore, at 500 °C the conversion into ceramic as well as the development of free carbon already started. The lower temperature of the transformation should be due to the lower heating rate and the presence of dwelling applied during annealing in comparison to the TGA. Nevertheless, the transformation is not completed and N=C=N groups are still present in all samples after annealing at 500 °C.

The interpretation of the fluorescence spectra of heat-treated polysilylcarbodiimides is not easy, since it is not possible to know the exact structure of the polymers evolved during the heat-treatment process. However, some considerations can be extrapolated from Figure 92, considering the different thermal behavior of the polymers. The four polysilylcarbodiimides analyzed can be distinguished in two categories: the polymers which crosslink during the temperature treatment up to 400 °C and those which do not. The polymers S1 and S2, possessing only methyl and phenyl substituents, are unable to crosslink during heating at temperatures lower than 400 °C. As attested by FT-IR and Raman analyses, no relevant changes were detected in the heat-treated polymers up to 400 °C (Figures 96 and 97). Thermal analysis and optical observations of the samples indicate the occurrence of the degradation of these polymers at 500 °C and no relevant weight losses are detected at lower temperatures (Figure 99). The crosslinking of S2 starts to occur at about 500 °C, when the methyl groups react and CH_4 is released. In S1 no crosslinking takes place even at higher temperatures. On the contrary, the polymers S3 and S4 start to crosslink at temperatures as low as 200 °C, due to the presence of Si-H bonds and vinyl groups, respectively, as attested by FT-IR and Raman analyses (Figures 96 and 97). As the TGA shows, S3 and S4 display more intense mass losses up to 400 °C than those of S1 and S2 (Figure 99). In S3, the weight loss is due to the hydrogen elimination correlated with the possible dehydrocoupling reactions, besides the end groups evolution [Morcos2008b]. In S4, the vinyl polymerization does not lead to mass loss, and the relatively high weight loss should be related to a higher extent of end groups evolution. In all polymers, the mass losses detected from 500 °C are due to the loss of H_2, phenyl derivatives, chlorine derivatives and C_2N_2, related to the polymer-ceramic transformation.

Considering the thermal behavior of the four polysilylcarbodiimides, we observe once again the fluorescence spectra. The polymers that crosslink during annealing (S3 and S4) are characterized by bathochromic shift of the maximum emission spectra as the treatment temperature increases, visible in Figure 92 (c) and (d). A red-shift of the maximum emission spectra with increasing annealing temperature was also detected in the silicon-based polymers examined in the previous chapters, where both the polysiloxane and the polysilazanes were able to crosslink.

In the case of the non-crosslinkable polymers S1 and S2, no bathochromic shift was found (Figure 92 (a) and (b)). The emission spectra of S1 remain unvaried up to 400 °C. Moreover, at this temperature the phenyl emission still prevails but a second luminescent center arises at 465 nm. In S2, except for the sample heat-treated at 300 °C, the maximum emission and excitation peaks correspond to the phenyl excimer emission up to 400 °C. After heat-treatment at 300 °C, the mentioned peak is only a shoulder and another luminescence center develops, which disappears after annealing at 400 °C. At 400 °C, a new weak peak develops at around 453 nm.

The photoluminescence analysis of the four annealed polysilylcarbodiimides is useful in order to better understand the luminescence mechanisms going on during heat-treatment of silicon-based polymers. In previous chapters, we deduced that the mechanisms responsible for the development of the luminescence properties of the polymers during heat-treatment, i.e. the rise of new red-shifted peaks, were mainly due to two reasons: the increased polymer-crosslinking and the development of free carbon. The presence of dangling bonds was initially also considered as source of luminescence, but EPR measurements on MK and Ceraset demonstrated that they do not have an appreciable influence on the development of the luminescence properties.

In this chapter, the role of the crosslinking was clearly envisaged. The presence of the phenyl groups provides a defined fluorescence emission in the polymers, which can be monitored with temperature. In polymer S1, no significant changes in the phenyl emission were detected after annealing up to 400 °C. In polymer S2, the maximum emission and excitation peaks correspond to the phenyl excimer transition at room temperature and after annealing at 200 °C. After heat-treatment at 300 °C, the mentioned peak is only a shoulder and other red-shifted emission peaks are detectable, which disappear after annealing at 400 °C. The luminescent centers that originate the red-shifted peaks are still under investigation and could be related to new π-π interactions of phenyl groups, which were identified by means of ^{1}H liquid state NMR and were not present in the starting polymers. In polymers S3 and S4, crosslinking is obviously responsible for the red-shift of the maximum intensity fluorescence spectra. In S3, crosslinking reactions through the Si-H bonds are assumed to be responsible for the observed red-shift in the fluorescence spectra. The interaction between two phenyl groups through the silicon atoms could lead to the formation of the new luminescent centers. In S4, the vinyl polymerization forms carbon containing bridges between silicon atoms, which can improve the π-π interactions of phenyl groups, causing red-shifted emission spectra as the annealing temperature increases.

However, the explanations supplied in the present chapter still cannot clarify the fluorescence mechanisms related to the crosslinking observed in previously analyzed polymers, where no phenyl groups were present as source of luminescence, and the mere crosslinking reactions developed new luminescence centers.

The role of free carbon on the luminescence properties starts to be detectable at higher temperatures, as in previously analyzed polymers. After heat-treatment at 500 °C, the fluorescence emission is quenched and red-shifted for all polysilylcarbodiimides. In S1, at 500 °C the luminescence is quenched to a weak emission with maximum peak at 645 nm, in S3 the maximum emission peak is at 574 nm, in S4 at 540 nm, while in S2 the fluorescence emission is completely quenched and no spectrum was collected. Thus, as previously found for the polysiloxane and polysilazane, the development of free carbon in higher amounts acts as a luminescence quencher.

^{13}C solid state MAS NMR spectra of the samples S1 and S2 annealed at 400 °C with heating rate 100 °C/h did not show the presence of free carbon, while after heat-treatment at 600 °C free carbon is detectable [Mera2009b]. For S3 and S4, solid state NMR is not yet available. In the samples annealed at 500 °C, Raman measurements show interference due to the heating of carbon clusters, which means that the free carbon should be present in high amounts. Moreover, from fluorescence analysis, optical observations (black color) and thermal analysis, we can attest that free carbon is developed in significant amounts in all polymers at a temperature as low as 500 °C.

Below that temperature, we assume the formation of small aromatic agglomerations that cannot be detected by means of NMR, which are the precursors of the free carbon sp^2 phase. As previously explained for MK, Ceraset and VL20, small aromatic agglomerations could be the origin of the luminescence emission mainly for annealing starting from 500 °C. Here, the structural changes occur at

6. Results and discussion

lower temperatures in respect to the previously analyzed polymers, as attested by the luminescence quenching at 500 °C. Therefore, free carbon could be present in low amounts after annealing of the polysilylcarbodiimides at 400 °C, although undetectable using solid state NMR and Raman analyses [Mera2009b]. Accordingly, in the case of MK at 500 and 600 °C and Ceraset at 500 °C, free carbon was not detected in solid state MAS NMR; however, sp^2 carbon radicals were detected in EPR measurements and C=C bonds in Raman for Ceraset at 500 °C. Therefore, the temperature of sp^2 carbon formation is lower than that detected by means of MAS NMR. Moreover, in the polysilylcarbodiimides the presence of phenyl groups could favor the formation of aromatic agglomerations at even lower temperatures. In MK and Ceraset, the formation of a low amount of sp^2 free carbon was concomitant with the yellow coloration of the samples and the visible photoluminescence. Here, the optical observations are not so useful, since the polysilylcarbodiimides are yellowish from lower treatment temperatures. Nevertheless, the samples assume a darker yellow coloration after treatment at 400 °C.

In order to differentiate between free carbon and crosslinking as the responsible features contributing to the analyzed luminescence, the photoluminescence spectra of polymers S1 and S2 must be taken into account first. After annealing at 400 °C, a shoulder at 465 nm arises in the maximum emission spectrum relative to S1 and a weak peak at 473 nm in S2. These peaks should be related to newly developed luminescent centers. As at 500 °C free carbon is present in significant amounts, it is plausible that at 400 °C some traces of free carbon, in form of small aromatic agglomerations, already started to form, which could be responsible for the observed peaks. Also in S3 and S4 the emission from aromatic agglomerations could cause new red-shifted peaks at heat-treatments from 400 °C, as in S1 and S2, but for these two polymers the analysis is complicated by the effect of crosslinking and it is difficult to distinguish among the different mechanisms, as previously for MK, Ceraset and VL20.

Anyhow, due to the systematical comparison between two polymers that crosslink and two that do not crosslink during annealing, it was possible to prove the role of the crosslinking on the fluorescence emission range. Moreover, the role of free carbon is evident in samples where crosslinking does not occur. Therefore, the control of the extent of crosslinking and amount of aromatic carbon agglomerations via the annealing temperature could be a useful tool to tune the emission range of the Si-based polymers for applications such as LEDs.

6.4. Copolymers

6.4.1. Polydiphenylsilylcarbodiimide + Polysilazane VL20

A new copolymer, GV, generated from the reaction of the polydiphenylsilylcarbodiimide (S1) with the polysilazane VL20 is presented in this chapter. The two starting polymers were mixed in equal weight proportions under stirring for one day at room temperature. It was not possible to mix the polymers in equal molar proportion because the molar weight of the polydiphenylsilylcarbodiimide is not known. The polymer was initially synthesized to attest the miscibility between the two different systems, but was analyzed in this work because of its fascinating fluorescence properties. In this study, we could show that the reaction of S1 with VL20 results in a copolymer. The first part of this section is focused on the investigation of the molecular structure of the newly synthesized GV polymer. Afterwards, the exceptional photoluminescence properties of the heat-treated GV samples will be presented, and the origin of the fluorescence will be studied by analyzing the structural changes during annealing.

6.4.1.1. Molecular structure

6.4.1.1.1. FT-IR spectroscopy

In Figure 103 the FT-IR structural analysis of the GV polymer is compared to the FT-IR of its precursors, S1 and VL20. All samples were measured using the ATR device. In Table 27 the vibration bands and respective bonds are listed.

6. Results and discussion

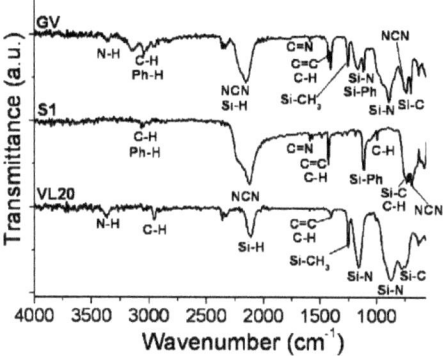

Figure 103: FT-IR analysis of the VL20, S1 and GV polymers.

Table 27: Vibration bands and respective bonds detected by means of FT-IR analysis of the GV samples.

Bond assignment	Vibration band (cm^{-1})
$\nu(N-H)$	3370
$\nu(OH)$	3151
$\nu(C-H\ sp^2,\ Ph-H)$	3050
$\nu_{as}(-CH_3)$	2948
$\nu(Si-H)+\nu_{as}(NCN)$	2158
$\nu_s(NCN)+\nu(C=C)$	1591
$\nu(C=C)$	1429
$\delta_{as}(C-H)$	1410
$\delta_s(Si-CH_3)$	1256
$\nu(Si-N-Si)$	1170
$\nu_s(Si-Ph)$	1121
$\nu_{as}(N-Si-N)+\nu(Si-N)$	894
$\delta_s(Si-C)+\gamma(NCN)$	739
$\nu(Si-C)$	698

The GV copolymer shows all the features of both starting polymers, by means of FT-IR. The band denoting OH bonds is also detectable.

6.4.1.1.2. Raman spectroscopy

The GV polymer was characterized by means of confocal micro-Raman spectrometer, with excitation laser of 488 nm. In Figure 104, the Raman spectrum of GV was plotted together with those relative to VL20 and S1, for comparison. In Table 28 the vibration bands and respective bonds are listed.

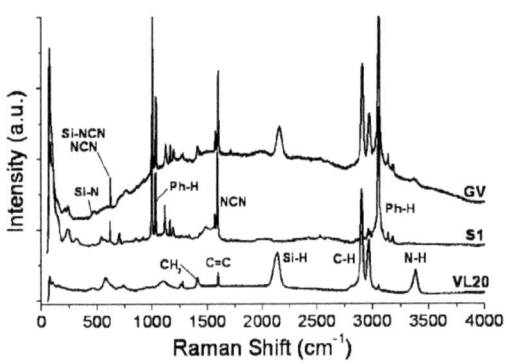

Figure 104: Raman spectra of VL20, S1 and GV polymers.

Table 28: Vibration bands and respective bonds detected by means of Raman analysis of the GV samples.

Bond assignment	Raman shift (cm^{-1})
$v(N-H)$	3379
$v(Ph-H)$	3051
$v_{as}(-CH_3)$	2968
$v_s(CH_2)$	2901/2905
$v(Si-H)$	2136/2154
$v_s(NCN)+v(C=C)$	1593/1587
$\delta_{as}(C-H)$	1406
$\delta_{wagg}(CH_2)$	1190
$\rho_{rock}(CH_2)$	1158
$\delta_{ip}(Ph-H)$	1026/1028

$\delta_{ip}(Ph\text{-}H)$	994/998
$v_{as}(Si(NCN))+\gamma(NCN))$	614
$v(Si\text{-}N)$	470

In the Raman spectrum of the copolymer it is possible to individuate all the bands of the starting polymers. The N-H band is significantly decreased in intensity in GV relatively to VL20.

6.4.1.1.3. XRD measurements

In Figure 105, the XRD diffractogram of the GV polymer is shown plotted together with those relative to VL20 and S1, for comparison.

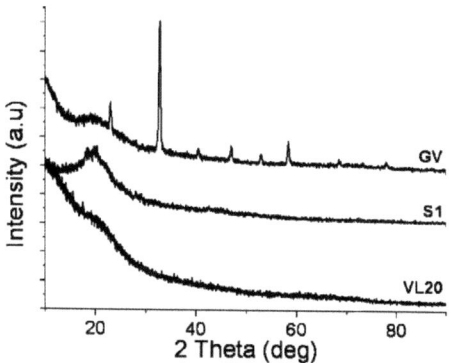

Figure 105: XRD spectra of the VL20, S1 and GV polymers.

Unlike the reacting polymers VL20 and S1, the copolymer GV shows a crystalline phase. This was identified as NH_4Cl having CsCl type structure with cubic lattice parameter 3.88 Å, which was confirmed by the Rietveld structure refinement of the XRD data (program Fullprof). The crystalline NH_4Cl can be highlighted under polarized microscope, which reveals small crystallites surrounded by the amorphous polymeric matrix.

6.4.1.1.4. Liquid State NMR spectroscopy (^{1}H, ^{13}C and ^{29}Si)

The reaction between the polydiphenylsilylcarbodiimide S1 and the polysilazane VL20 produced a new copolymer. FT-IR and Raman analyses demonstrated that GV is characterized by features of both the starting polymers, but they could not attest that a reaction occurred. X-ray diffraction highlighted the formation of a new crystalline phase, the NH$_4$Cl. In Figure 106, the ^{29}Si DEPT (a), the ^{13}C{^{1}H} (b) and the ^{1}H (c) NMR spectra obtained in C$_6$D$_6$ of GV, S1 and VL20 are presented.

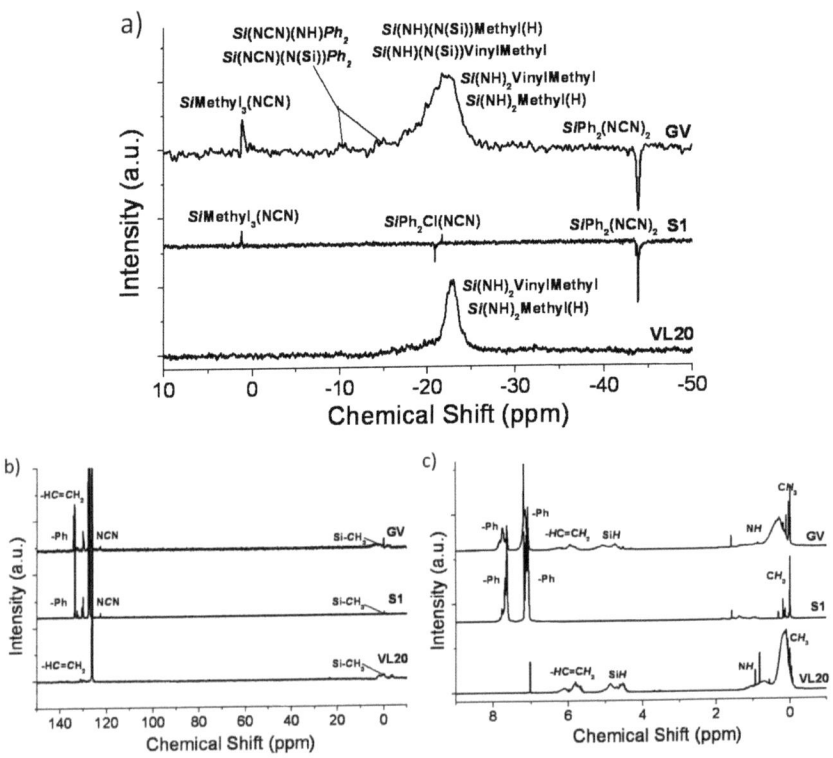

Figure 106: ^{29}Si DEPT (a), ^{13}C{^{1}H} (b), ^{1}H (c) NMR spectra of the VL20, S1 and GV polymers in C$_6$D$_6$.

The ^{29}Si DEPT NMR spectrum of the GV copolymer is generally defined by the sum of the chemical shifts present in the NMR spectra of the reacting polymers.

In the ^{29}Si DEPT NMR (C_6D_6), the Si(NH)$_2$Methyl(H) and Si(NH)$_2$VinylMethyl signals at -22.7 ppm from the VL20 are present also in the copolymer, although significantly broadened. Also the SiPh$_2$(NCN)$_2$ at -43.7 ppm, typical of the S1 chain, and the end group SiMethyl$_3$(NCN) at 1.2 ppm are found in GV. Nevertheless, the signal relative to the end group SiPh$_2$Cl(NCN), present in S1 at -20.8 ppm, is not detectable in GV.

Besides the peaks previously present in S1 and VL20, it is possible to identify two additional signals at -9.9 ppm and at -14.9 ppm, present only in the final polymer. Since the NCN group shows in NMR the same chemical shift of oxygen [Riedel1998, Mera2009b], and the O-Si-NH chemical shift is well-matched with -15 ppm [Imbenotte1979], the new peaks at -9.9 and -14.9 ppm could indicate the presence of Si bonded both to the amine and the carbodiimide groups, the NCN-Si-NH group. Besides the NCN-Si-N(H) groups, relative to the linear chains (see discussion, Figure 110 (c)), the peaks can also be assigned to NCN-Si-N(Si), relative to the ramified structure (Figure 110 (b)), both formed by the reaction. The reaction route and the structures formed will be explained in the discussion section. Furthermore, the broadening of the peak at -22.7 ppm, better explained with the formation of additional peaks in the region -16.8 to -22.2 ppm, could be due to the ramification of the polysilazane through the transamination reaction (see discussion, Figure 110 (a)).

In ^{13}C{^{1}H} and ^{1}H NMR the spectra relative to the GV consist of the sum of the signals present in the starting polymers and no additional peaks were detected.

In ^{13}C NMR (C_6D_6), the GV polymer shows the methyl groups (at -4 to 2 ppm) from both VL20 and S1 (end groups), vinyl groups (at 126-131 ppm) from the VL20 and the NCN (at 122 ppm) and phenyl (at 127-133 ppm) groups from the S1.

In ^{1}H NMR (C_6D_6), the methyl groups (at 0-2 ppm) from both VL20 and S1, the vinyl (at 5-7 ppm), Si-H (at 4-5 ppm) and N-H (at 0.8-1.2 ppm) groups from the VL20, as well as the phenyl groups (at 7-8 ppm) from the S1 are detectable in the GV polymer.

Therefore, liquid state NMR analysis of GV demonstrates that a reaction occurred between VL20 and S1 and a copolymer was generated. Furthermore, the chlorine from the end groups of S1 took part to the reaction, as attested by the disappearance in GV of the signal relative to SiPh$_2$Cl(NCN) at -20.8 ppm in ^{29}Si NMR.

6.4.1.1.5. Discussion on the molecular structure

In this section, we will discuss the possible reactions between the polymers VL20 and S1 and propose a molecular structure for the GV polymer on the basis of the spectroscopic characterizations performed.

The reaction between VL20 and S1 generates a crystalline by-product, namely NH_4Cl, identified by means of XRD (Figure 105). The ammonia present in the ammonium chloride derives from the NH groups of the polysilazane chains and the Cl from the end groups of the polydiphenylsilylcarbodiimide. Ammonia is formed from the polysilazane via transamination reactions, as illustrated in Figure 107 [Birot1995].

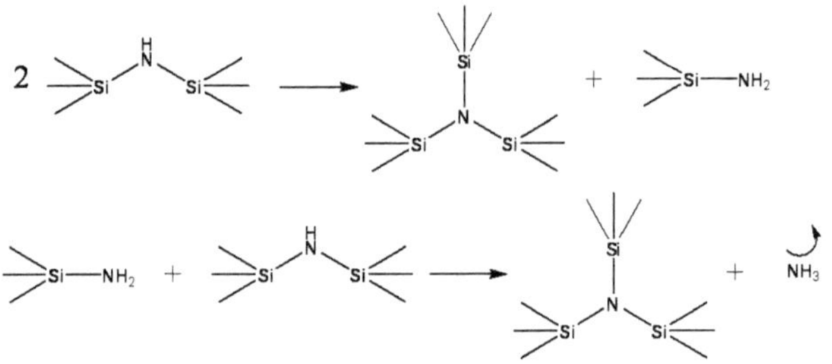

Figure 107: Transamination reaction.

The effect of the transamination reaction is, besides the development of ammonia, also the formation of ramified polysilazane chains, with nitrogen bonded to three silicon atoms (Figure 110 (a)).
The chlorine from the polydiphenylsilylcarbodiimide end groups, in proximity of ammonia and the NH group from a polysilazane chain, reacts to generate NH_4Cl (Figure 108). The reaction occurs via formation of HCl, which subsequently reacts with the NH_3.

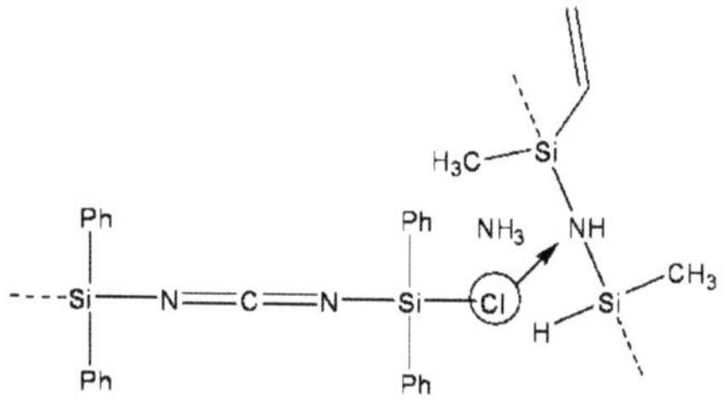

6. Results and discussion

Figure 108: Reaction between the polydiphenylsilylcarbodiimide end group and the polysilazane chain.

The silicon atom can bond to the nitrogen from the polysilazane chain, forming a hybrid ramified structure, composed by two polysilazane chains and a polydiphenylsilylcarbodiimide chain (Figure 110 (b)).

Moreover, the polydiphenylsilylcarbodiimide end groups can also react with the ≡Si-NH$_2$ end groups formed by transamination reactions (Figure 109): HCl is released, which can react with the ammonia and form NH$_4$Cl, and a linear hybrid polymer is obtained (Figure 110 (c)).

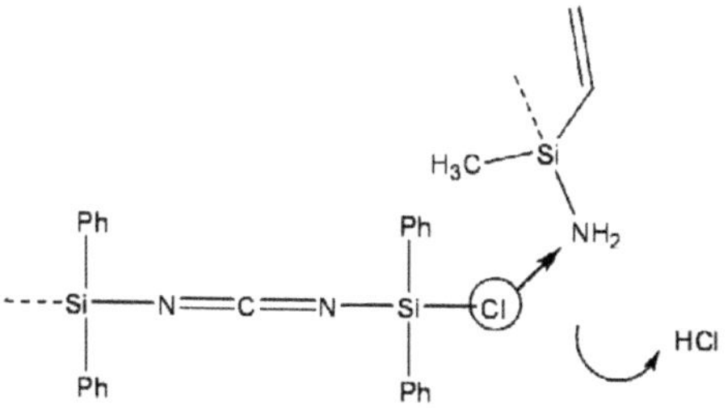

Figure 109: Reaction between the polydiphenylsilylcarbodiimide end group and the ≡Si-NH$_2$ end group formed by the transamination reaction.

Through the last two reactions, the N(Si)-Si-NCN and N(H)-Si-NCN species (silicon bonded both to the nitrogen and to the carbodiimide group) are formed, as attested by liquid state ^{29}Si NMR analysis (Figure 106). In Figure 110, the possible structures present in the GV polymer, formed by the previously explained reactions, are shown.

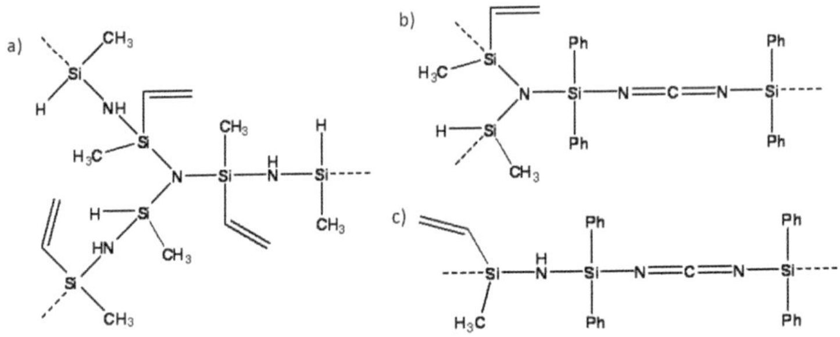

Figure 110: Possible structural units present in the GV polymer.

The reaction between the chlorine from the end groups of the polydiphenylsilylcarbodiimide and the polysilazane chain, with formation of HCl, is probably the first step of the total reaction and it is the cause of the transamination reaction, which otherwise would not occur in the polysilazane.

Other reactions possibilities were not confirmed by spectroscopic analyses. For example, the Si-H groups of the VL20 are present after the reaction of the polymers, as attested from Raman, FT-IR and ^{1}H NMR spectroscopy (Figure 104, 103 and 106). In Raman, the N-H band is significantly decreased in intensity in the GV polymer in comparison to VL20 and this is due to the formation of the crystalline structure NH_4Cl. Liquid state NMR, FT-IR and Raman analyses show that the structures of the starting polymers are preserved after the reaction and GV is a copolymer composed by the combination of the building blocks of the starting polymers.

The polymer GV possesses the crosslinkable substituents of the VL20, the vinyl, Si-H and N-H groups, which can react through hydrosilylation, vinyl polymerization, dehydrocoupling and transamination reactions. As the VL20 is composed by a mixture of low molecular weight oligomers and high molecular weight polymeric chains, the polymer GV should also have this feature.

The GV polymer was heat-treated at 200, 300, 400, 500 and 600 °C for 2 h under Ar flow, with heating rate 50 °C/h and free cooling. Crosslinked porous samples were obtained, which were subsequently powdered.

6.4.1.2. Photoluminescence measurements

6.4.1.2.1. Fluorescence measurements

The fluorescence emission and excitation spectra of GV and the heat-treated samples, relative to the maximum intensity excitation and emission wavelengths (Table 29), are displayed in Figure 111.

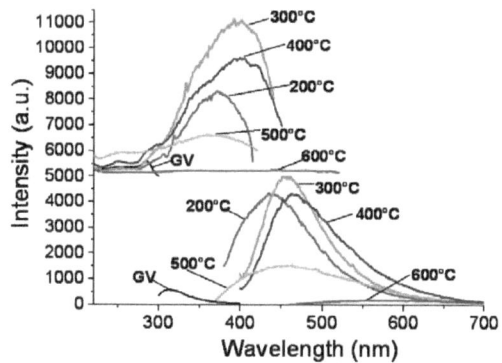

Figure 111: Fluorescence emission (bottom right) and excitation (top left) spectra of GV annealed at different temperatures, obtained with excitation and emission wavelengths according to Table 29.

Table 29: Excitation and emission wavelengths used for the GV samples.

Sample	Excitation wavelength (nm)	Emission wavelength (nm)
GV	285	313
GV 200 °C	369	435
GV 300 °C	394	454
GV 400 °C	393	461
GV 500 °C	360	445
GV 600 °C	446	547

As in the case of its precursor VL20, blue and blue-greenish emission was obtained for heat-treated GV samples up to 500 °C. After heat-treatment at 200 °C, the maximum fluorescence emission is considerably red-shifted in respect to that of the untreated polymer. A slight red-shift of the emission maxima with increasing temperature could be detected in the spectra of samples heat-treated between 200 and 400 °C. The sample annealed at 500 °C shows quenched and not red-shifted emission. After heat-treatment at 600 °C, a weak green emission is detectable, which denotes a considerable decrease in emission intensity and the red-shift of the emission peak.

The GV polymer displays maximum emission at 313 nm after excitation at 285 nm.

After heat-treatment at 200 °C, maximum emission with the main peak at around 443 nm and two additional smaller peaks at 397 nm and 528 nm is detected with an excitation wavelength of 369 nm.

The sample heat-treated at 300 °C shows maximum emission after excitation at 394 nm, and the emission peaks are at 457 nm and 544 nm.

After heat-treatment at 400 °C, with excitation at 393 nm, emission peaks at 464 nm and 544 nm were detected.

After heat-treatment at 500 °C, the whole emission intensity is decreased and the maximum emission is obtained for excitation at 360 nm with emission peaks at 400 nm, 451 nm and 532 nm.

Finally, after heat-treatment at 600 °C, a large emission band with maximum at 579 nm is obtained with excitation at 446 nm.

The maximum excitation spectra of the heat-treated GV samples show two main bands. The main excitation peak for all samples is between 360 and 380 nm and a second peak at 250 nm is detectable.

In Figure 112 optical pictures of the polymer GV after annealing at different temperatures are shown under white (a) and UV (360-400 nm) (b) light, respectively.

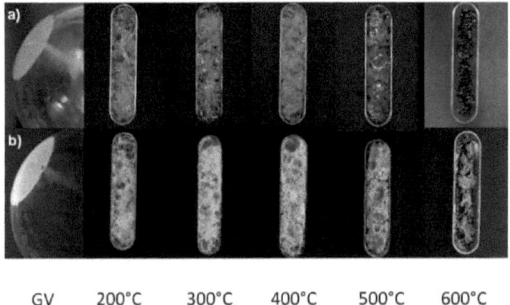

Figure 112: Photographs representing GV samples under white (a) and UV (360-400 nm) (b) light. For colored pictures the reader can refer to http://tuprints.ulb.tu-darmstadt.de/2085/.

Since the excitation wavelengths to obtain maximum emission are in the range 360-400 nm for all samples, GV is an interesting candidate for LED applications. In addition, the emission intensities are much higher in comparison to those detected in previously analyzed samples, as shown by the high intensity values in Figure 111 and by the optical pictures in Figure 112. In Figure 113, the emission spectra of the GV samples obtained with excitation wavelength of 360 nm are illustrated.

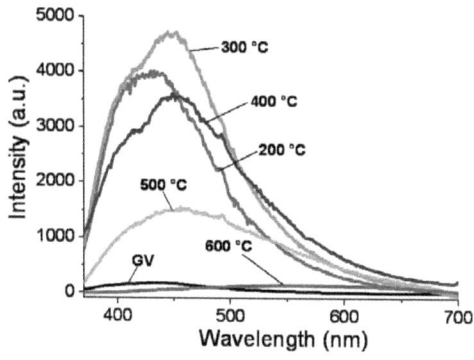

Figure 113: Fluorescence emission spectra of GV annealed at different temperatures obtained with 360 nm as excitation wavelength.

The emission curves of the different samples obtained with excitation wavelength of 360 nm show the presence of three common peaks, two main peaks at 395-400 nm and 443-451 nm and a smaller band at 525-561 nm. The GV polymer, after excitation at 360 nm presents an emission peak at 438 nm. After heat-treatment at 200 °C, the main peak is at around 443 nm, with two additional smaller bands at 400 nm and 528 nm. Following annealing at 300 °C, the maximum peak is at 451 nm, with less intense peaks at 395 nm and 544 nm. After heat-treatment at 400 °C, the same emission peaks at 448, 395 and 525 nm are present, with decreased intensity. After heat-treatment at 500 °C the whole emission intensity is decreased and three peaks are detectable at 400 nm, 451 nm and 561 nm. Finally, after heat-treatment at 600 °C, a large and weak emission band composed of two peaks at 561 nm and 458 nm is obtained.

The emission spectra of the GV samples obtained with excitation wavelength of 250 nm are shown in Figure 114.

6. Results and discussion

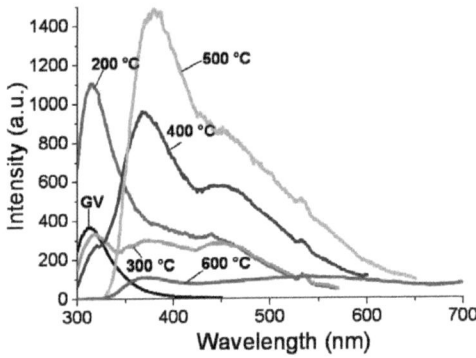

Figure 114: Fluorescence emission spectra of GV annealed at different temperatures obtained with 250 nm as excitation wavelength.

With 360 nm as the excitation wavelength, the emission ranges and intensities appear similar for the samples annealed at 200, 300 and 400 °C. With 250 nm as the excitation wavelength, it is possible to observe a consistent transformation of the emission spectra with the annealing temperature. Emission peaks at 314 nm, 368-379 nm, 450 nm and 544 nm are detected in all heat-treated samples, having different relative intensity ratios, which changes as the heating temperature increases. The polymer GV shows one emission peak, at 314 nm. After annealing at 200 °C, the more intense emission band is still at 316 nm, but other two less intense peaks are detectable at 375 nm and 450 nm. After annealing at 300 °C, the whole emission intensity is reduced and the peaks present at 316 nm, 375 nm and 450 nm have more or less the same intensity. After heat-treatment at 400 °C, the peak at 316 nm becomes a shoulder, while the maximum emission peak is at 368 nm; the peak at 450 nm increases in intensity and a new peak at 544 nm forms a shoulder. After heat-treatment at 500 °C, the emission intensity with excitation at 250 nm increases further; the main peak is at 379 nm, the shoulder at 316 nm disappears, while the shoulders at 451 nm and 544 nm are still present. After annealing at 600 °C, two peaks are detectable at 375 nm and 544 nm.

As general observation, the emission intensity with excitation at 250 nm is considerably reduced compared to that obtained with excitation at 360 nm. The maximum emission intensity with excitation at 250 nm is found after annealing at 500 °C. Increasing the annealing temperature of the precursor causes an increased relative intensity of the longer-wavelength peaks and a decreased intensity of the shorter-wavelength peaks. Following annealing at 600 °C the emission is drastically reduced.

In order to explain the fluorescence spectra obtained with excitation wavelengths 250 and 360 nm, the maximum excitation spectra of the GV samples of Figure 111 must be considered. The excitation peaks are relative to the different emission maxima of every sample; however, some common considerations can be extrapolated. The main excitation peak for all samples is between 360 and 400 nm and a second peak at 250 nm is present. Therefore, with 360 nm as excitation wavelength, more intense spectra were obtained. The band at 250 nm shows low intensity in all samples up to 300 °C. Accordingly, after excitation at 250 nm, weak emissions were obtained at around 450 nm (the excitation spectra were obtained in average in relation to this emission wavelength). After treatment from 400 °C, the excitation peak at 250 nm gains relative importance in comparison to the peak at 360-400 nm. Accordingly, after excitation at 250 nm, red-shifted emission spectra are obtained in respect to lower temperatures, which contribute to the maximum emission spectra at around 450 nm.

With 250 nm as the excitation wavelength, the phenyl groups are excited. Emission at 313 nm is obtained for the GV precursor. As the treatment temperature increases, new red-shifted emission peaks are detectable. They are related to new luminescence centers, probably due to π-π interactions between phenyl groups, caused by the increasing crosslinking of the polymer. With increasing annealing temperature, the red-shifted peaks due to the π-π interactions acquire importance in respect to the original peak.

On the contrary, the excitation peak at 360 nm, with emission in the range 443-451 nm, should be related to a new luminescence center, created by the crosslinking reactions, and probably not associated to the phenyl emission.

The spectral emissions obtained with 360 nm excitation of the GV samples heat-treated at 200-600 °C are represented on the CIE chromaticity diagram 1931 (Figure 115).

6. Results and discussion

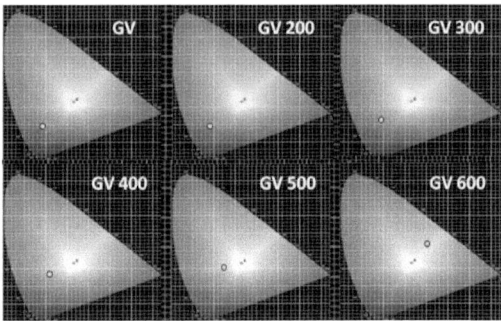

Figure 115: CIE chromaticity diagrams 1931 of the GV samples annealed at 200-600 °C. For colored pictures the reader can refer to http://tuprints.ulb.tu-darmstadt.de/2085/.

The color coordinates are perfectly in agreement with the colors of the samples under UV light (Figure 112).

6.4.1.2.2. Stability of the fluorescence during storage

In order to monitor the change in fluorescence properties of the GV samples following storage in air, the fluorescence emission obtained with 250 nm excitation wavelength was measured during the first week after the heat-treatment and about one year later (Figure 116).

6. Results and discussion

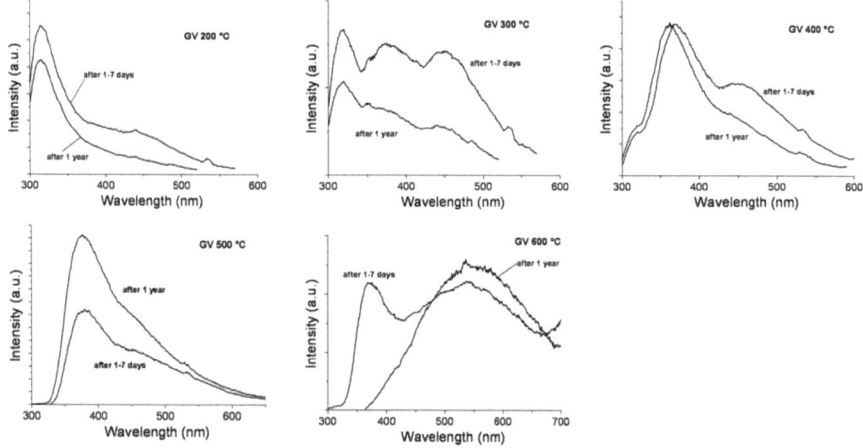

Figure 116: Fluorescence spectra of the GV samples obtained during the first week after heat-treatment and about 1 year later (at 250 nm excitation).

After heat-treatment at 200, 300, 400 and 500 °C the changes are mainly in the emission intensity, remaining the peaks unvaried. For the samples heat-treated up to 400 °C, the emission intensity decreases after 1 year in air, while the sample heat-treated at 500 °C shows an intensity increase. The sample heat-treated at 600 °C does not show intensity changes, but the peak at 375 nm disappears and the only detectable peak after 1 year in air is at 544 nm.

6.4.1.2.3. Quantum efficiency

The quantum efficiency of the sample GV 300 °C was measured with the integrating sphere and resulted to be 22%. This is the highest value found among the heat-treated silicon-based polymers analyzed in the present thesis. It is an encouraging result, because it is much higher than the values found for the starting materials, and attests that the fluorescence properties of the silicon-based polymers can be enhanced.

The quantum efficiencies of the remaining GV samples were estimated by comparing the integral of their emission spectra (in energy scale) with the integral of the emission spectra of samples of known quantum efficiency with emission in the same range (Table 30).

Table 30: Quantum efficiencies of the GV samples.

Sample	Quantum efficiency (%)
GV	6.3*
GV 200 °C	29.5
GV 300 °C	22.0
GV 400 °C	21.1
GV 500 °C	16.2
GV 600 °C	1.0

The samples designed with * emit in the UV range, for which no measurement was successful. Thus the values were calculated by comparison with a sample with emission maximum at 394 nm. Since the emission range of GV 200 °C was intermediate between those of Ceraset 500 °C and GV 300 °C, its quantum efficiency was calculated considering both samples. A very high value was obtained for this sample, higher than that of GV 300 °C, in accordance with the larger integral of its spectrum.

6.4.1.3. Absorption measurements

Additional information about the electronic transitions occurring in the GV samples and their variation after temperature treatment can be obtained with absorption measurements.

6.4.1.3.1. UV-Vis-NIR spectroscopy

In Figure 117, the absorption spectrum of the GV polymer obtained with UV-Vis spectroscopy is compared to those relative to the two starting polymers S1 and VL20.

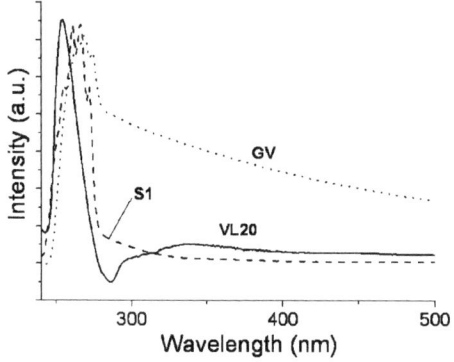

Figure 117: UV-Vis absorption spectra of GV, VL20 and S1.

The absorption edge of GV at around 280 nm (4.43 eV) is similar to those of S1 (around 280 nm (4.43 eV)) and VL20 (around 285 nm (4.35 eV)).

As previously attested for other silicon-based polymers, the solubility of GV decreases as the treatment temperature increases. However, standard UV-Vis measurements could be performed on the GV samples heat-treated up to 400 °C, in THF solution, as previously for the S1 samples (Figure 118).

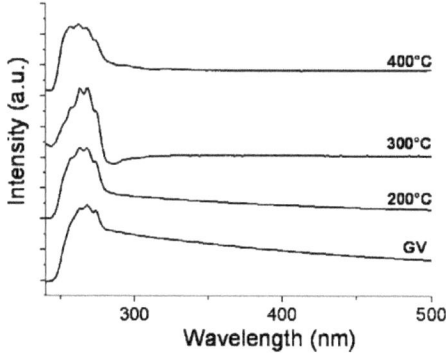

Figure 118: UV-Vis absorption spectra of GV samples.

The absorption edge is at around 280 nm up to 400 °C. Vibrational structure is present in all spectra, as in the polydiphenylsilylcarbodiimide, and denotes the absorption from phenyl groups. A new red-shifted weak peak is detectable at 294 nm in the sample heat-treated at 400 °C, which could be attributed to the absorption from aromatic agglomerations.

6.4.1.3.2. Reflection measurements

The reflection spectra were obtained using synchronous excitation and detection wavelengths with the Cary Eclipse Varian. The absorption/scattering spectra obtained from the reflection data relative to the GV samples (using the Kubelka-Munk function) are presented in Figure 119. The spectra were smoothed for representative reasons. The intensities of GV 200 °C and especially GV 500 °C were significantly higher than those of the remaining samples, therefore they were also normalized.

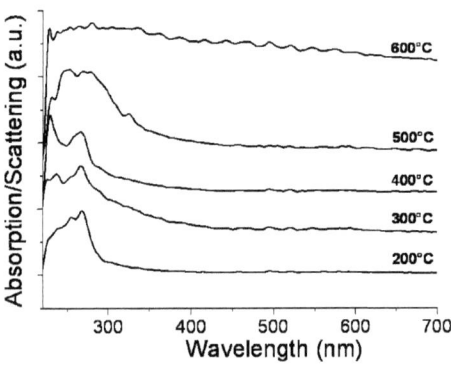

Figure 119: Absorption/scattering spectra relative to the GV samples obtained with reflection measurements.

The absorption range of the GV samples increases toward less energetic wavelength as the treatment temperature increases, as in the previously analyzed samples. The sample annealed at 200 °C absorbs radiation up to about 400 nm (3.10 eV), after heat-treatment at 300 and 400 °C up to about 450 nm (2.76 eV), after heat-treatment at 500 °C the absorption edge is at about 480 nm (2.58 eV), while at 600 °C the sample absorbs in the whole visible range. The absorption edges are in accordance with the colors of the samples and with the excitation spectra of Figure 111. In GV samples, the red-shift of the absorption edge is not so marked as in previously analyzed samples, and this is in accordance with photoluminescence measurements.

6.4.1.4. FT-IR spectroscopy

The GV and the powdered samples heat-treated at 200, 300, 400, 500 and 600 °C were studied with FT-IR structural analysis (Figure 120). All samples were measured using the ATR device.

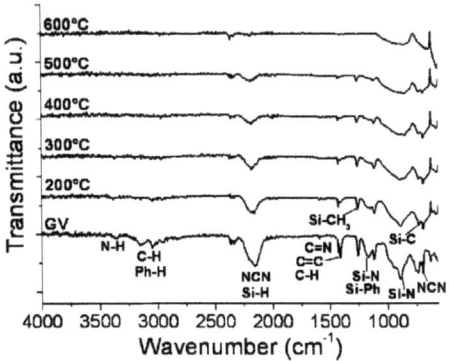

Figure 120: FT-IR analysis of the GV samples.

The assignment of the bands was already performed for the untreated copolymer (Table 27).
The C=C band decreases in temperature and demonstrates that crosslinking occurs also through vinyl groups. The N-H and C-H bands, as well as the Si-H/N=C=N band, decrease in intensity with increasing annealing temperature, suggesting the occurrence of dehydrocoupling and hydrosilylation reactions. At 600 °C, N=C=N and C=N bonds are not detectable anymore. The Si-N and the Si-C bonds broaden as the annealing temperature increases, demonstrating the ceramization of the polymer.

6.4.1.4. Raman spectroscopy

The heat-treated GV samples were investigated with micro-Raman spectrometer with excitation laser of 488 nm, but high fluorescence interferences were detected. Therefore, they were also analyzed with the IR/Raman spectrometer Bruker IFS 55 - FRA 106, with laser wavelength 1064 nm. The spectra are reported in Figure 121.

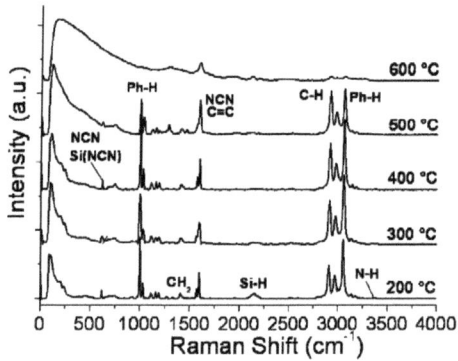

Figure 121: Raman spectra of the GV samples heat-treated at different temperatures using 1064 nm as laser wavelength.

The assignment of the bands was already performed for the untreated copolymer (Table 28).
The sample heat-treated at 600 °C shows interference, attributed to the heating of the carbon clusters. The Ph-H, C-H, C=C/NCN and Si-NCN bonds appear unvaried up to 500 °C. The Si-H peak is visible up to 300 °C, but it disappears at higher treatment temperatures, indicating the occurrence of crosslinking reactions through Si-H groups. The N-H peak is only slightly detectable. The peaks at 1285 and 1460 cm^{-1}, indicating unsaturated C-H groups and C=C bonds, are detectable at 200 °C and considerably increase in intensity at 500 °C, due to the formation of free carbon.

6.4.1.5. TGA/DTG/MS

Thermal analysis (TGA/DTG/MS) was useful to monitor the thermal behavior of the GV polymer (Figure 122).

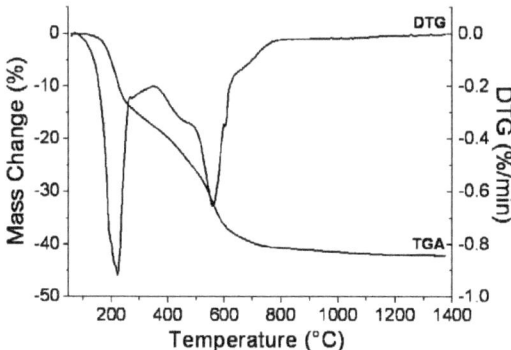

Figure 122: TGA and DTG of the GV polymer.

Selected MS spectra of gaseous byproducts relative to the thermal transformation of GV are shown in Figure 123.

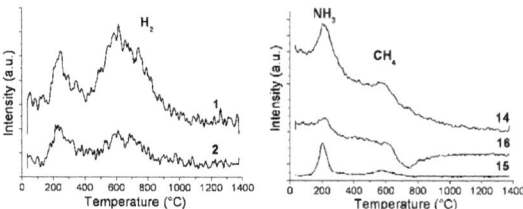

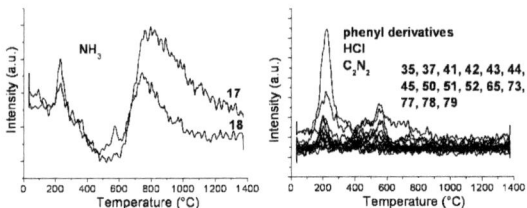

Figure 123: Selected TG/MS analysis of the gaseous byproducts of the thermal decomposition of GV. The numbers in the graphs refer to m/z.

The TGA of GV shows three main mass losses, the first between 200 and 300 °C, due to the loss of NH_3, H_2 and oligomers (not clearly detectable by MS analysis) from the VL20 and phenyl derivatives from the S1 polymer. The NH_4Cl, individuated in the polymer by XRD analysis, degrades upon heating in NH_3 and HCl, which also leave the polymer below 300 °C. The second and third mass losses, between 350 and 800 °C, are due to the elimination of H_2, CH_4, NH_3, C_2N_2 and phenyl derivatives.

The TGA/DTG graphs relative to the starting polymers (VL20 (a) and S1 (b)) and to GV (c) are displayed in Figure 124 for comparison.

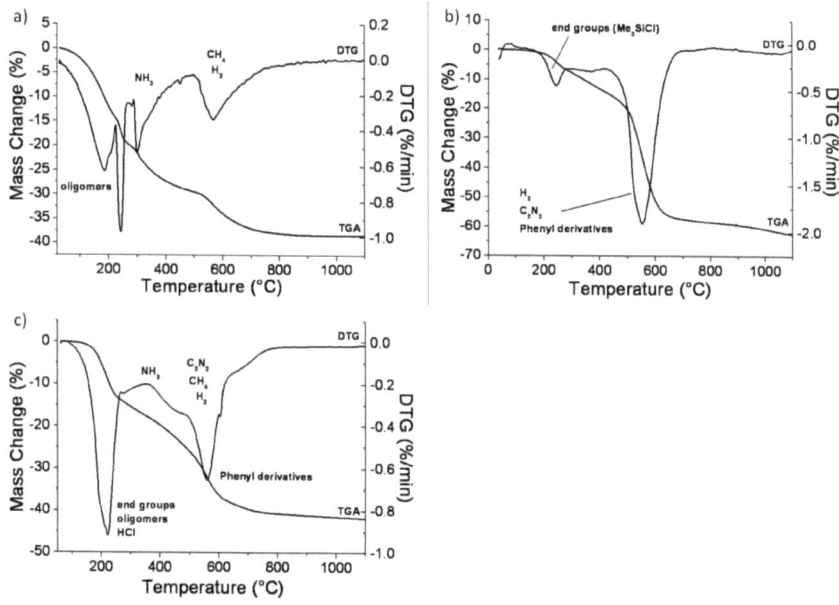

Figure 124: Thermal analyses (TGA and DTG) of the VL20 (a), polydiphenylsilylcarbodiimide (b) and GV (c).

In the TGA/DTG curve of VL20 four mass losses could be identified: up to 220 °C, due to oligomers losses; between 220 and 450 °C, relative to the elimination of NH_3, and between 500 and 600 °C, relative to the elimination of CH_4 and H_2 (Figure 124 (a)), as reported in chapter 5.2.2.5. The TGA curve of polydiphenylsilylcarbodiimide shows two mass losses, between 200 and 300 °C, relative to the elimination of end groups (Me_3SiCl), and between 500 and 600 °C, due to the removal of phenyl derivatives, C_2N_2 and H_2 (Figure 124 (b)), as reported in chapter 5.3.1.5. The thermal analysis relative to the GV polymer (Figure 124 (c)) shows that the species removed generally correspond to the sum of the species eliminated in the separate starting polymers. One difference is that during thermolysis of the GV polymer, HCl is eliminated during the first mass loss step, due to the degradation of the crystalline phase NH_4Cl, while in the TG/MS of S1 Me_3SiCl species were detected.

6.4.1.6. XRD measurements

In Figure 125, the X-ray diffractograms of the GV polymer and of the annealed samples are shown.

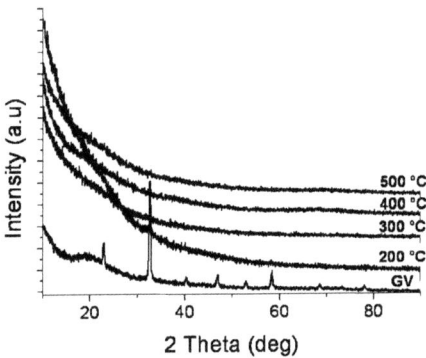

Figure 125: X-ray diffraction patterns of the GV samples.

The crystalline phase, identified as NH₄Cl and present in the GV copolymer, disappears after heat-treatment. Its decomposition compounds are NH_3 and HCl, detected by means of TG/MS. After heat-treatment at 200 °C the most intensive reflex is still visible.

6.4.1.7. Multinuclear Solid State MAS NMR spectroscopy (^{1}H, ^{13}C and ^{29}Si)

^{1}H, ^{13}C CP and ^{29}Si MAS NMR spectra of GV heat-treated at 200, 300, 400, 500 and 600 °C are shown in Figure 126 and their chemical shifts and bonds attribution is listed in Table 31. The spinning side bands are labeled with * and were not taken into account in the simulations.

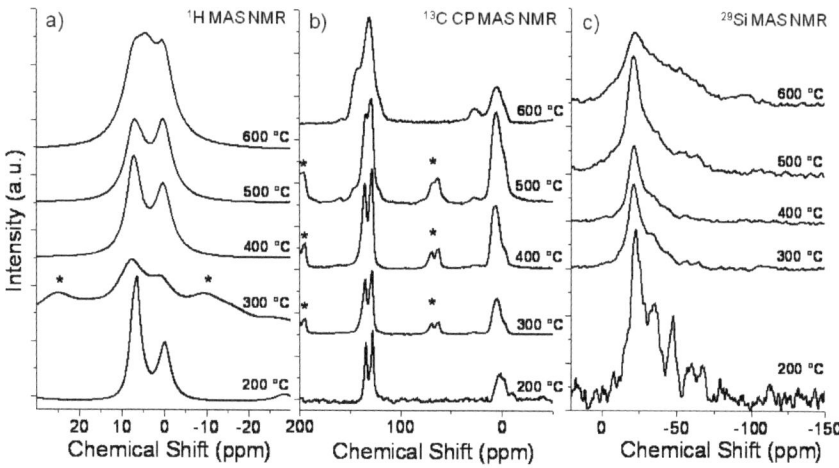

Figure 126: a) ^{1}H, b) ^{13}C CP and c) ^{29}Si MAS NMR spectra of GV heat-treated at different temperatures.

Table 31: Chemical shifts and relative bonds detected in MAS NMR spectra for GV samples heat-treated at different temperatures.

	^{1}H MAS NMR	^{13}C CP MAS NMR	^{29}Si MAS NMR
GV 200 °C	0.0 ppm (CH_3, CH_2); 6.5, 7.6 ppm (protonated sp^2 carbon (-HC$_{arom}$))	-9.3 ppm (C-Si-NCN); -2.2, 2.0, 5.3 ppm (SiCH_3, Si-CH_2-CH_2-Si); 127.9, 134.4 ppm (aromatic C)	-60.0 ppm (Si-O); -47.6 ppm ((Ph)$_2$Si(NCN)$_2$, Si-O); -35.1 ppm (SiCN$_3$, Si-O); -23.1 ppm (SiHCN$_2$, SiC$_2$N$_2$)
GV 300 °C	0.5 ppm (CH_3, CH_2); 7.9 ppm (protonated sp^2 carbon (-HC$_{arom}$))	-2.2, 5.1 ppm (SiCH_3, Si-CH_2-CH_2-Si); 128.6, 135.4 ppm (aromatic C)	-21.6 ppm (SiC$_2$N$_2$); -33.6 ppm (SiCN$_3$)
GV 400 °C	0.3 ppm (CH_3, CH_2); 7.2 ppm (protonated sp^2 carbon (-HC$_{arom}$))	-2.6, 4.5, 8.6 ppm (SiCH_3, Si-CH_2-CH_2-Si); 128.7, 135.6 ppm (aromatic C)	-21.8 ppm (SiC$_2$N$_2$); -32.7 ppm (SiCN$_3$); -42.5 ppm ((Ph)$_2$Si(NCN)$_2$, SiN$_4$)
GV 500 °C	0.3 ppm (CH_3, CH_2); 6.9 ppm (protonated sp^2 carbon (-HC$_{arom}$))	-2.0, 4.3, 8.4 ppm (SiCH_3, Si-CH_2-CH_2-Si); 121.7 ppm (N=C=N); 128.9	-21.3 ppm (SiC$_2$N$_2$); -30.3 ppm (SiCN$_3$); -59.6 ppm (SiN$_4$, (Ph)$_2$Si(NCN)$_2$)

		and 134.6 ppm (aromatic C); 146.6 ppm (Csp^2, Csp^2-N)	
GV 600 °C	0.2 ppm (CH_3, CH_2); 4.2 ppm (adsorbed water, unsaturated CH_2);7.5 ppm (protonated sp^2 carbon (-HC$_{arom}$))	5.0 ppm (SiCH_3, Si-CH_2-CH_2-Si); 27.3 ppm (C-C); 121.2 ppm (N=C=N); 131.2 ppm (aromatic C); 143.0 ppm (Csp^2, Csp^2-N)	-22.0 ppm (SiC$_2$N$_2$); -36.5 ppm (SiCN$_3$, (Ph)$_2$Si(NCN)$_2$); -54.6 ppm (SiN$_4$, (Ph)$_2$Si(NCN)$_2$)

6.4.1.7.1. ^{1}H MAS NMR

The ^{1}H MAS NMR spectra of GV heat-treated from 200 to 600 °C (Figure 126 (a)) show the presence of CH_3 and CH_2 bonds at around 0 ppm and of protonated carbon sp^2 (-HC$_{arom}$) at around 6-8 ppm. At 600 °C, a signal at 4-5 ppm indicates adsorbed water or unsaturated CH_2.

6.4.1.7.2. ^{13}C CP MAS NMR

The ^{13}C MAS NMR spectrum of GV heat-treated at 200 °C (Figure 126 (b)) shows carbon bonded to silicon that is bonded to a NCN unit (C-Si-NCN) at -9.3 ppm, Si-CH$_3$ or Si-CH$_2$-CH$_2$-Si bonds at -2.2, 2.0, 5.3 ppm and aromatic carbon at 127.9 and 134.4 ppm. From 300 to 600 °C, SiCH$_3$ or Si-CH$_2$-CH$_2$-Si signals are detected at -3 to 9 ppm with broadening of the signal with increasing temperature and aromatic carbon is detected at around 128 and 135 ppm. At 500 and 600 °C, a new signal denoting free carbon (partially as carbon sp^2 bonded to nitrogen, Csp^2-N) is visible at around 145 ppm. At 600 °C, a peak denoting C-C bonds is detectable at 27.3 ppm. The signal at around 121 ppm indicates N=C=N groups and is detectable in the samples annealed at 500 and 600 °C.

6.4.1.7.3. ^{29}Si MAS NMR

The ^{29}Si MAS NMR spectrum of GV after heat-treatment at 200 °C (Figure 126 (c)) shows the presence of [SiC$_2$N$_2$] or [SiHCN$_2$] units at -23.1 ppm; the peak at -35.1 ppm is attributed to [SiCN$_3$] or to Si-O due to oxidation, the peak at -47.6 ppm to [(Ph)$_2$$Si(NCN)_2$] or to Si-O due to oxidation, and the peak at -60 ppm indicates the Si-O bonds due to oxidation related to the long storage in air before the NMR measurements. After treatment at 300 °C, the oxidation is not discernible anymore. The peak at -21.6 ppm indicates [SiC$_2$N$_2$], while the one at -33.6 ppm [SiCN$_3$] units. No peak can be related to (Ph)$_2$$Si(NCN)_2$. From 400 to 600 °C, the [SiC$_2$N$_2$] unit is observable at around -22 ppm, the [SiCN$_3$] unit from -30 to -36 ppm, the [SiN$_4$] from -42 to -59 ppm. The peaks at -42.5 for GV 400 °C, at -59.6 ppm for GV 500 °C and at -36.5 and -54.6 ppm for GV 600 °C are also partially due to [(Ph)$_2$Si(NCN)$_2$]. The peaks in the range -10 to -15 ppm, corresponding to silicon bonded to both NCN and NH or N(Si) groups and detected in liquid state ^{29}Si NMR, were not identified by simulation of the solid state NMR spectra, due to their proximity and low intensity compared to the main peak.

6.4.1.7.4. Discussion of the Solid State MAS NMR results

From ^{1}H and ^{13}C MAS NMR, the free carbon starts to be detected in GV at a temperature as low as 500 °C. GV polymer heat-treated at 200 °C seems to be completely crosslinked, but it is not easy to discern in ^{1}H and ^{13}C NMR the peaks relative to N-H, Si-H and vinyl groups, because they are hidden by the signal relative to -HC$_{arom}$ from phenyl groups. From ^{29}Si NMR, up to 300 °C, the [SiC$_2$N$_2$] (at 200 °C it could also be [SiHCN$_2$]) and [SiCN$_3$] units are visible, while after heat-treatments from 400 °C also the [SiN$_4$] unit is appreciable. Air contamination due to the long storage in air before the measurements is clearly visible in the sample annealed at 200 °C, but not in the samples heat-treated at higher temperatures.

6.4.1.8. Discussion

The GV polymer is interesting because of the intense fluorescence emission shown after heat-treatment. Both the starting polymers show photoluminescence properties after heat-treatment, but the copolymer obtained from their reaction, after heat-treatment, presents enhanced fluorescence intensities. The photoluminescence and q.e. measurements of the GV samples revealed the highest fluorescence intensity among all previously characterized silicon-based polymers (Figure 111 and Table 30).
As in the case of the previously discussed silicon-based polymers, it is necessary to take into account several mechanisms in order to explain the development of the luminescence properties of the copolymer GV after heat-treatment. With increasing temperature, the separation of free sp^2 carbon and the luminescence centers generated by the crosslinking reactions have an influence on the formation of new red-shifted fluorescence peaks. The formation of radicals was observed not to have an appreciable effect on the development of the photoluminescence properties. Moreover, phenyl groups are present in GV.
In the synthesized GV copolymer, the fluorescence should be exclusively caused by phenyl emission (Figure 127).

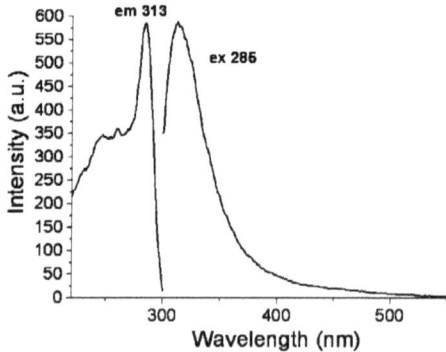

Figure 127: Fluorescence excitation and emission spectra obtained with emission wavelength of 313 nm and excitation wavelength of 285 nm, respectively.

The maximum emission of the GV polymer is at 313 nm and is less intense than the S1 emission band in the range 328-348 nm (Figure 88). The maximum excitation spectrum presents for both polymers maximum at 285-300 nm and some less intense peaks at higher energies. Nevertheless, the emission spectrum of GV does not present the vibrational structure detected in S1. The absence of vibrational structure and the blue-shift of the emission compared to S1 indicate the modification of the delocalized unit Ph-SiX$_2$-Ph. In S1, this unit is characterized by an angle between the phenyl groups that is maximized (180°), while in GV it probably modifies into a less planar system, due to the constraint of a ramified polymer and to the presence of the VL20.

The GV polymer shows emission intensity comparable to those of the VL20 and S1 precursors. Nevertheless, the heat-treated GV samples display emission intensities much higher than those of the starting polymers, heat-treated or not. Therefore, during annealing of the GV polymer, a new luminescence center with intense fluorescence properties must develop. The improved intensity and the red-shift of the fluorescence excitation and emission peaks of the heat-treated GV samples in comparison to the GV polymer should be correlated to a luminescence center that develops through the crosslinking reactions. The formation of sp^2 free carbon, which starts at relatively higher temperatures, could create further red-shifted excitation and emission peaks.

As previously mentioned, the excitation peak at 250 nm in Figure 111 is related to the phenyl transition. The emission spectra obtained with 250 nm as excitation wavelength change after heat-treatment, showing the formation of new red-shifted peaks (Figure 114). They are to attribute to new π-π interactions between the phenyl groups, which form due to the crosslinking of the polymer. With increasing treatment temperature, the more red-shifted peaks acquire importance. The emission due to phenyl groups does not show intensity improvements after heat-treatment.

The excitation peak at 360 nm in Figure 111, corresponding to the emission in the range 443-451 nm, is not related to the fluorescence from phenyl groups, but it should be related to a new luminescent center that develops by means of crosslinking. This luminescence center is slightly detectable also in the precursor, with excitation at 360 nm, but after annealing its emission intensity is significantly enhanced. Thus, the improved emission of the annealed samples should be correlated to the crosslinking of the GV polymer. After annealing at 200 °C, when the crosslinking starts, the maximum emission is red-shifted and much more efficient in comparison to the phenyl emission of the GV polymer. After treatment of the GV at 300 and 400 °C, the crosslinking continues and further small red-shifts are observable. As in the case of MK, Ceraset and VL20, the luminescence center created through the crosslinking has not been yet identified. Future studies will be focused on this issue.

From ^{1}H and ^{13}C MAS NMR, the free carbon starts to be detected in GV at 500 °C (Figure 126). As previously explained, it is plausible that the formation of free carbon starts at lower temperatures than that detected by means of NMR, in the case of GV at 400 °C. The decreased temperature of free carbon formation in GV compared to VL20 is related to the presence of polydiphenylsilylcarbodiimide, which is rich in carbon and contains phenyl groups, and for which free carbon is also assumed to start forming at 400 °C. At this temperature, small aromatic agglomerations could generate new red-shifted emission peaks. Nevertheless, at 400 °C this emission is probably too weak in comparison with the emission from the luminescent center generated by crosslinking reactions and it is in the same wavelength range. From the photoluminescence spectra of Figure 111, the free carbon formation at heat-treatment temperatures from 400 °C seems to counteract the excitation band at 360-400 nm. At 500 and 600 °C, the further development of free carbon (detectable in MAS NMR) leads to a decrease of the emission intensities. After annealing at 500 °C, no red-shift of the maximum emission spectrum is observable, as the detected emission is still due to the luminescence center generated by the crosslinking, although decreased in intensity. At 600 °C, the photoluminescence is quenched because of the occurrence of mineralization reactions and further free carbon development: a weak red-shifted emission is observed, which is to attribute to the free carbon.

In summary, GV shows intriguing visible fluorescence when thermally treated. The quantum efficiency measured for the sample heat-treated at 300 °C was equal to 22±5%. The possibility of q.e. improvement renders this polymer, and the silicon-based polymers in general, very interesting for luminescence application, where high efficiency is required.

Besides the high emission intensity, also the emission and excitation ranges have interesting features. The maximum excitation peaks of the GV copolymer heat-treated between 200 and 500 °C correspond to the range 360-400 nm, which is the required excitation range in LED applications. The maximum emission bands are characterized by broad spectra, completely in the visible range.

As mentioned before, the maximum emission intensities are obtained for annealing at 200, 300 and 400 °C. Such low treatment temperatures permit the direct molding and subsequent annealing of the polymer in LED devices.

Nevertheless, as the photographs in Figure 112 show, in the specific case of the GV copolymer the application for LEDs could be hindered by the release of gases during annealing which remain trapped

6. Results and discussion

in the polymer, with creation of macroscopic porosity. Heat-treatment under vacuum does not solve the problem. The trapping of gases was not observed at the same extent for the heat-treated starting polymers. TGA analysis shows that at 300 °C the weight loss in S1 is 10%, in VL20 about 22%, while in GV about 16%. At 500 °C the weight loss in S1 is 20%, in VL20 30%, in GV 28%. Thus, in the GV copolymer, the high mass loss due to the release of species from the VL20 combined with an increased viscosity from the S1, due to the presence of longer chains and steric phenyl groups, and a ramified structure, lead to the production of gases whose release from the material is hindered during annealing.

Moreover, from the viewpoint of LED applications, the fluorescence stability in air is also an issue, as in the other silicon-based polymers; however, application in protective atmosphere could be a suitable alternative solution.

In conclusion, the copolymer composed of a polysilazane and a polysilylcarbodiimide offers an important step forward in the novel field of luminescent heat-treated silicon based polymers, because it demonstrates the possibility of creation of new luminescent centers by means of annealing through crosslinking reactions, which offer high quantum efficiency and emission in the visible range, obtained with low annealing temperatures.

6.5. References

[An2004] L. An, W. Xu, S. Rajagopalan, C. Wang, H. Wang, Y. Fan, L. Zhang, D. Jiang, J. Kapat, L. Chow, B. Guo, J. Liang, R. Vaidyanathan, *Adv. Mater.*, 16, *22*, 2004

[Andrade2006] A. A. Andrade, M. F. Coutinho, M. P. P. de Castro, H. Vargas, J.H. Rohling, A. Novatski, N. G. C. Astrath, J. R. D. Pereira, A. C. Bento, M. L. Baesso, S. L. Oliveira, L.A.O. Nunes, *J. Non-Cryst. Solids,* 352, *3624–3627*, 2006

[Andronenko2006] S. I. Andronenko, I. Stihau, S. K. Misra , *J. Appl. Phys.*, 99, *113907*, 2006

[Anggono2006] J. Anggono, B. Derby, *J. Eur. Ceram. Soc.*, 26, *1107–1119*, 2006

[Asthana2006] A. Asthana, Y. Asthana, I. K. Sung, D. P. Kim, *Lab. Chip,* 6, *1200–1204*, 2006

[Bahadur2003] D. Bahadur, D. Das, *Proc. Indian Acad. Sci. (Chem. Sci.*), 115 [5-6], *587–606*, 2003

[Bakumov2007] V. Bakumov, K. Gueinzius, C. Hermann, M. Schwarz, E. Kroke, *J. Eur. Ceram. Soc.*, 27, *3287–3292*, 2007

[Balan2006]C. Balan, R. Riedel, *J. Optoelectron. Adv. M.*, 8[2], *561 – 567*, 2006

[Balci2005] M. Balci, *Basic ^1H- and ^{13}C-NMR Spectroscopy*, Ch. 13, Jan-2005, Imprint: ELSEVIER

[Baran2004] M. Baran, B. Bulakh, N. Korsunska, L. Khomenkova, J. Jedrezejewski, *Eur. Phys. Appl. Phys.*, 27, *285-287*, 2004

[Berger2005] F. Berger, A. Müller, F. Aldinger, K. Z. Müller, *Anorg. Allg. Chem.*, 631, *355*, 2005

[Bernardo2006] E. Bernardo and P. Colombo S. Hampshire, *J. Am. Ceram. Soc.*, 89 [12], *3839–3842*, 2006

[Bernardo2007] E. Bernardo, *J. Eur. Ceram. Soc.*, 27, *2415–2422*, 2007

[Bernardo2009] E. Bernardo, P. Colombo, S. Hampshire, *J. Eur. Ceram. Soc.*, 29, *843–849*, 2009

[Birot1995] M. Birot, J.-P. Pillot, J. Dunoguès, *Chem. Rev.*, 95, *1443-1477*, 1995

[Bitterlich2005] B. Bitterlich, J. G. Heinrich, *J. Am. Ceram. Soc.*, 88 [10], *2713–2721*, 2005

[Bois1994] L. Bois, J. Maquet, F. Babonneau, H. Mutin, D. Bahloul, *Chem. Mater.*, 6, *796-802*, 1994

[Bolbit2000] N. M. Bolbit, V. B. Taraban, E. R. Klinshpont, I. P. Shelukhov, V. K. Milinchuk, *High Energ. Chem.*, 34 [4], *229-235*, 2000

[Bothelho do Rego2001] A. M. Bothelho do Rego, O. Pellegrino, M. Rei Vilar, *Macromolecules*, 34, *4987-4992*, 2001

[Brahmandam2007] S. Brahmandam, R. Raj, *J. Am. Ceram. Soc.*, 90 [10], *3171–3176*, 2007

[Colombo2002] P. Colombo, J. R. Hellmann, *Mat. Res. Innovat.*, 6, *260–272*, 2002

[Colombo2003] P. Colombo, K. Perini, E. Bernardo, T. Capelletti, G. Maccagnan, *J. Am. Ceram. Soc.*, 86 [6], *1025–27*, 2003

[Colombo2008] P. Colombo, *J. Eur. Ceram. Soc.*, 28, *1389–1395*, 2008
[Cooke2003] D. W. Cooke, R. E. Muenchausen, B. L. Bennett, D. A. Wrobleski, E. B. Orler, *Radiat. Phys. Chem.*, 66, *129–135*, 2003
[Costacurta2007] S. Costacurta, L. Biasetto, E. Pippel, J. Woltersdorf, P. Colombo, *J. Am. Ceram. Soc.*, 90 [7], *2172–2177*, 2007
[Cramer2004] N. B. Cramer, S. Reddy, H. Lu, T. Cross, R. Raj, C. N. Bowman, *J. Polym. Sci. A1*, 42, *1752–1757*, 2004
[Cross2006a] T. J. Cross, R. Raj, S. V. Prasad, D. R. Tallant, *Int. J. Appl. Ceram. Technol.*, 3 [2], *113–126*, 2006
[Cross2006b] T. J. Cross, R. Raj, T. J. Cross, S. V. Prasad, D. R. Tallant, 2006, *The American Ceramic Society*, 3 [2], 2006
[Czerw2001] R. Czerw, Z. Guo, P. M. Ajayan, Y.-P. Sun, D. L. Carroll, *Nano Lett.*, 1 [8], *423-427*, 2001
[Das2004] D. Das, A. Saha, C. M. Srivastava, R. Raj, S. E. Russek, D. Bahadur, *J. Appl. Phys.*, 95 [11], 2004
[Dhamne2005] A. Dhamne, W. Xu, B. G. Fookes, Y. Fan, L. Zhang, S. Burton, J. Hu, J. Ford, L. An, *J. Am. Ceram. Soc.*, 88 [9], *2415–2419*, 2005
[Dias2000] F. B. Dias, J. C. Lima, A. l. Maçanita, S. J. Clarson, A. Horta, I. F. Piérola, *Macromolecules*, 33, *4772-4779*, 2000
[Dresselhaus2000] M. S. Dresselhaus, M. A. Pimenta, P. C. Eklund, G. Dresselhaus, *Raman scattering in fullrenes and related carbon-based materials* in *Raman Scattering in Materials Science*, Springer Series in Materials Science, vol 42, Wber WH, Merlin R (eds). Anita Publications: New Dehli, 2002, 392
[Duan2004] R.-G. Duan, J. D. Kuntz, J. E. Garay, A. K. Mukherjee, *Scripta Mater.*, 50, *1309–1313*, 2004
[Duan2005] R.-G. Duan, A. K. Mukherjee, *Scripta Mater.*, 53, *1071–1075*, 2005
[Eremenko1969] A. M. Eremenko, E. F. Sheka, M. A. Piontkovskaya, I. E. Neimark, *Teoreticheskaya i Eksperimental'naya Khimiya*, 5 [2], *242-246*, 1969
[Ferraioli2008] L. Ferraioli, D. Ahn, A. Saha, L. Pavesi, R. Raj, *J. Am. Ceram. Soc.*, 91 [7], 2422–2424, 2008
[Friedel1966] R. A. Friedel, H. L. Gibson, *Nature*, 211 [5047], *404-405*, 1966
[Friedel2005] T. Friedel, N. Travitzky, F. Niebling, M. Scheffler, P. Greil, *J. Eur. Ceram. Soc.*, 25, *193–197*, 2005
[Gadow2003] R. Gadow, F. Kern, *Adv. Eng. Mat.*, 5 [11], *799-801*, 2003
[Gao2008a] F. Gao, W. Yang, Y. Fan, L. An, *Nanotechnology*, 19, 105602, 2008
[Gao2008b] F. Gao, W. Yang, Y. Fan, L. An, *J. Solid State Chem.*, 181, *211–215*, 2008
[Garcia1995] J. Garcia M., M.A. Mondragon, C. Tellez S., A. Campero, V.M. Castano, *Mater. Chem. Phys.*, 41, *15-17*, 1995
[Garcia2003] C. B. W. Garcia, C. Lovell, C. Curry, M. Faught, Y. Zhang, U. Wiesner, *J. Polym. Sci. Pol. Phys.*, 41, *3346–3350*, 2003

[Gasch2001] M. J. Gasch, J. Wan, A.K. Mukherjee, *Scripta Mater.*, 45, *1063-1068*, 2001
[Green1997] W. H. Green, K. P. Le, J. Grey, T. T. Au, M. J. Sailor, *Science,* 276, *1826,* 1997
[Green2001] M. A. Green, J. Zhao, A. Wang, P. J. Reece, M. Gal, *Nature,* 412, *805-808*, 2001
[Gudapati2006] V. M. Gudapati, V. P. Veedu, M. N. Ghasemi-Nejhad, *Compos. Sci. Technol.,* 66, *3230–3240,* 2006
[Hanemann2002] T. Hanemann, M. Ade, M. Börner, G. Motz, M. Schulz, J. Haußelt, *Adv. Eng. Mater.,* 4, 11, *869-873,* 2002
[Harshe2004a] R. R. Harshe, C. Balan, R. Riedel, *J. Eur. Ceram. Soc.,* 24, *3471–3482,* 2004
[Harshe2004b] R.R. Harshe, *PhD thesis,* TU Darmstadt, 2004
[Hauser2008] R. Hauser, A. Francis, R. Theismann, R. Riedel, *J. Mater. Sci.,* 43, *4042–4049,* 2008
[Hayakawa2003] T. Hayakawa, A. Hiramitsu, M. Nogami, *Appl. Phys. Lett.,* 82, *18,* 2003
[Herzog2005] A. Herzog, M. Thünemann, U. Vogt, O. Beffort, *J. Eur. Ceram. Soc.,* 25, *187–192,* 2005
[Imbenotte1979] M. Imbenotte, G. Palavit, M. L. Filleux-Blanchard, *Z. Anorg. Allg. Chem.,* 455, *103-111*, 1979
[Ionescu2009b] E. Ionescu, C. Linck, C. Fasel, M. Müller, H.-J. Kleebe, R. Riedel, *J. Am. Ceram. Soc.,* 93 [1], *241–250,* 2010
[Ischenko2006] V. Ischenko, R. Harshe, R. Riedel, J. Woltersdorf, *J. Organomet. Chem.,* 691, *4086–4091,* 2006
[Janakiraman2006] N. Janakiraman, T. Höche, J. Grins, S. Esmaeilzadeh, *J. Mater. Chem.,* 16, *3844–3853,* 2006
[Janakiraman2009] N. Janakiraman, F. Aldinger, *J. Eur. Ceram. Soc.,* 29, *163–173,* 2009
[Jones2009] B. H. Jones, T. P. Lodge, *J. Am. Chem. Soc.,* 131 [5], *1676-1677,* 2009
[Kamperman2004] M. Kamperman, C. B. W. Garcia, P. Du, H. Ow, U. Wiesner, *J. Am. Chem. Soc.,* 126 [45], *14708-14709,* 2004
[Kamperman2007] M. Kamperman, P. Du, R. O. Scarlat, E. Herz, U. Werner-Zwanziger, R. Graf, J. W. Zwanziger, H. W. Spiess, U. Wiesner, *Macromol. Chem. Phys.,* 208, *2096–2108,* 2007
[Kamperman2008] M. Kamperman, M. A. Fierke, C. B. W. Garcia, U. Wiesner, *Macromolecules,* 41 [22], *8745-8752,* 2008
[Katsuda2006] Y. Katsuda, P. Gerstel, J. Narayanan, J. Bill, F. Aldinger, *J. Eur. Ceram. Soc.,* 26, *3399–3405,* 2006
[Kim1997] H. K. Kim, M.-K. Ryu, S.-M. Lee, *Macromolecules,* 30, *1236-1239*, 1997
[Kim1998] H. K. Kim, M.-K. Ryu, K.-D. Kim, S.-M. Lee, S.-W. Cho, J.-W. Park, *Macromolecules,* 31, *1114-1123,* 1998
[Kim2007] B.-S. Kim, D. J. Kim, *J. Eur. Ceram. Soc.,* 27, *837–841,* 2007
[KiON] Technical Bulletin 1, KiON Ceraset polyureasilazane and KiON Ceraset polysilazane 20 heat-curable resins of KiON Corporation, http:// www.kioncorp.com/bulletins.html (Last accessed 03.09).
[Klaffke2006] D. Klaffke, R. Wäsche, N. Janakiraman, F. Aldinger, *Wear,* 260, 711–719, 2006

6. Results and discussion

[Kojima2002] A. Kojima, S. Hoshii, T. Muto, *J. Mater. Sci. Lett.*, 21, *757– 760*, 2002
[Kolb2006] R. Kolb, C. Fasel, V. Liebau-Kunzmann, R. Riedel, *J. Eur. Ceram. Soc.*, 26, *3903–3908*, 2006
[Larsen2006] K.L. Larsen, O. F. Nielsen, *J. Raman Spectrosc.*, 37, *217-222*, 2006
[Lee2005] S.-H. Lee, M. Weinmann, F. Aldinger, *J. Am. Ceram. Soc.*, 88 [11], *3024–3031*, 2005
[Lee2006] K.-S. Lee, *Adv. Funct. Mater.*, 16, *1235-1241*, 2006
[Lee2007] D.-H. Lee, K.-H. Park, L.-Y. Hong, D.-P. Kim, *Sensors Actuat. A*, 135, *895–901*, 2007
[Liebau-Kunzmann2006] V. Liebau-Kunzmann, C. Fasel, R. Kolb, R. Riedel, *J. Eur. Ceram. Soc.*, 26, *3897–3901*, 2006
[Liemersdorf2008] S. Liemersdorf, R. Riedel, J. Oberlez, *J. Am. Ceram. Soc.*, 91 [1], *325–328*, 2008
[Li2000a] Y.-L. Li, R. Riedel, J. Steiger, H. von Seggern, *Adv. Eng. Mater.*, 2 [5], 2000
[Li2001] Y. L. Li, E. Kroke, R. Riedel, C. Fasel, C. Gervais, F. Babonneau, *Appl. Organometal. Chem.*, 15, *820–832*, 2001
[Li2007] Y.-H. Li, X.-D. Li, D.-P. Kim, *J. Organomet. Chem.*, 692, *5303–5306*, 2007
[Li2008] Y. Li, L. Fernandez-Recio, P. Gerstel, V. Srot, P. A. van Aken, G. Kaiser, M. Burghard, J. Bill, *Chem. Mater.*, 20 [17], *5593-5599*, 2008
[Liew2001] L.-A. Liew, W. Zhang, V. M. Bright, M. L. Dunn, L. An, R. Raj, *Sensors Actuat. A*, 89, *64-70*, 2001
[Lin2000] J. Lin, K. Baerner, *Mater. Lett.*, 46, *86–92*, 2000
[Lin2005] Y. Lin, B. Zhou, R. B. Martin, K. B. Henbest, B. A. Harruff, J. E. Riggs, Z.-X. Guo, L. F. Allard, Y.-P. Sun, *J. Phys. Chem. B*, 109 [31], *14779-14782*, 2005
[Liu2002] Y. Liu, L.-A. Liew, R. Luo, L. An, M. L. Dunn, V. M. Bright, J. W. Daily, R. Raj, *Sensors Actuat. A*, 95, 143-151, 2002
[Mackin2000] T. J. Mackin, M, C. Roberts, *J. Am. Ceram. Soc.*, 83 [2], *337–343*, 2000
[Maire2007] E. Maire, P. Colombo, J. Adrien, L. Babout, L. Biasetto, *J. Eur. Ceram. Soc*, 27, *1973–1981*, 2007
[Markgraaff1996] J. Markgraaff, *J. S. Afr. I. Min. Metall.*, 55-65, 1996
[Menapace2008] I. Menapace, G. Mera, R. Riedel, E. Erdem, R.-A. Eichel, A. Pauletti, G. A. Appleby, *J. Mater. Sci.*, 43, *5790–5796*, 2008
[Mera2009b] G. Mera, R. Riedel, F. Poli, K. Müller, *J. Eur. Ceram. Soc*, 29, *2873-2883*, 2009
[Morcos2008a] R. M. Morcos, A. Navrotsky, T. Varga, D. Ahn, A. Saha, F. Poli, K. Müller, R. Raj, *J. Am. Ceram. Soc.*, 91 [7], *2391-2393*, 2008
[Morcos2008b] R. M. Morcos, G. Mera, A. Navrotsky, T. Varga, R. Riedel, F. Poli, K. Müller, *J. Am. Ceram. Soc.*, 91 [10], *3349–3354*, 2008
[Moysan2007] C. Moysan, R. Riedel, R. Harshe, T. Rouxel, F. Augereau, *J. Eur. Ceram. Soc*, 27, *397–403*, 2007

[Nagaiah2006] N. R. Nagaiah, A. K. Sleiti, S. Rodriguez, J. S. Kapat, L. An, L. Chow, *J. Phys. Conf. Ser.,* 34, *277–282*, 2006

[Nangrejo2009] M. Nangrejo, E. Bernardo, P. Colombo, U. Farook, Z. Ahmad, E. Stride, M. Edirisinghe, *Mater. Lett.*, 63, *483–485*, 2009

[Nghiem2007] Q. D. Nghiem, D. Kim, D.-P. Kim, *Adv. Mater.,* 19, *2351-2354*, 2007

[Ni2008] Z. Ni, Y. Wang, T. Yu, Z. Shen, *Nano Res.*, 1, *273 291*, 2008

[O'Connell2002] M. J. O'Connell, S. M. Bachilo, C. B. Huffman, V. C. Moore, M. S. Strano, E. H. Haroz, K. L. Rialon, P. J. Boul, W. H. Noon, C. Kittrell, J. Ma, R. H. Hauge, R. B. Weisman, R. E. Smalley, *Science,* 297, *593*, 2002

[Pham2006] T. A. Pham, D.-P. Kim, T.-W. Lim, S.-H. Park, D.Y. Yang, K.-S. Lee, *Adv. Funct. Mater.*, 16 [9], 1235-1241, 2006

[Pham2007] T. A. Pham, P. Kim, M. Kwak, K. Y. Suh, D.-P. Kim, *Chem. Commun.*, *4021–4023*, 2007

[Pivin1998] J. C. Pivin, P. Colombo, M. Sendova-Vassileva, J. Salomon, G. Sagon, A. Quaranta, *Nucl. Instrum. Meth. B,* 141, *652-662*, 1998

[Pivin2000] J.C. Pivin, M. Sendova-Vassileva, P. Colombo, A. Martucci, *Mater. Sci. Eng. B,* 69–70, *574–577*, 2000

[Plovnick2000] R. H. Plovnick, D. J. Pysher, *Mater. Res. Bull.,* 35, *1453–1461*, 2000

[Prasad2000] B. L. V. Prasad, H. Sato, T. Enoki, Y. Hishiyama, Y. Kaburagi, A. M. Rao, P. C. Eklund, K. Oshida, M. Endo, *Phys. Rev. B*, 62, *11209-11218*, 2000

[Prokes1998] S. M. Prokes, W. E. Carlos, S. Veprek, Ch. Ossadnik, *Phys. Rev. B*, 58, *23*, 1998

[Quimby1998] R. S. Quimby, P. A. Tick, N. F. Borrelli, L. K. Cornelius, *J. Appl. Phys.,* 83 [3], *1*, 1998

[Radovanovic1999] E. Radovanovic, M. Gozzi, M. Goncalves and I. Yoshida, *J. Non-Cryst. Solids*, 248, *37–48*, 1999

[Rak2001] Z. S. Rak, *J. Am. Ceram. Soc.,* 84 [10], *2235–2239*, 2001

[Rambo2006a] C. R. Rambo, H. Sieber, *J. Mater. Sci.,* 41, *3315–3322*, 2006

[Rambo2006b] C. R. Rambo, J. A. Junkes, H. Sieber, D. Hotza, *Exacta,* 4 [2], *297-308,* 2006

[Reddy2003] S. K. Reddy, N. B. Cramer, T. Cross, R. Raj, C. N. Bowman, *Chem. Mater.*, 15 [22], *4257-4261*, 2003

[Reddy2004] S. K. Reddy, N. B. Cramer, A. K. O´Brien, T. Cross, R. Raj, C. N. Bowman, *Macromol. Symp.* 206, *361–374*, 2004

[Riedel1998] R. Riedel, E. Kroke, A. Greiner, A. O. Gabriel, L. Ruwisch, J. Nicolich, *Chem. Mater.*, 10, *2964-2979*, 1998

[Rocha2005] R. M. Rocha, P. Greil, J. C. Bressiani, A. H. A. Bressiani, *Mater. Res.*, 8 [2], *191-196*, 2005

[Rocha2006] R.M. Rocha, C.A.A. Cairo, M.L.A. Graca, *Mater. Sci. Eng. A,* 437, *268–273*, 2006

[Rocha2008] R. M. Rocha, E. A. B. Moura, A. H. A. Bressiani, J. C. Bressiani, *J. Mater. Sci.*, 43, *4466–4474*, 2008

[Roh2008] C. Roh, S.-H. Lee, V. Francois, *J. Chromatogr. A*, 1179, *145–151*, 2008

[Rohwer2003] L. S. Rohwer, A. M. Srivastava, *Electrochem. Soc. Inteface,* 12 [2], *36-39*, 2003

[Ritzhaupt-Kleissl2006] H.-J. Ritzhaupt-Kleissl, J. R. Binder, T. Gietzelt, J. Kotschenreuther, *Adv. Eng. Mater.*, 8 [10], 2006

[Ryu2007] H.-Y. Ryu, R. Raj, *J. Am. Ceram. Soc.*, 90 [1], *295–297*, 2007

[Saha2003a] A. Saha, S. R. Shah, R. Raj, S. E. Russek, *J. Mater. Res.*, 18 [11], 2003

[Saha2003b] A. Saha, S. R. Shah, R. Raj, *J. Am. Ceram. Soc.*, 86 [8], *1443–1445*, 2003

[Salimgaareeva2003] V. N. Salimgareeva, S. S. Ostakhov, V. A. Ponomareva, S. V. Kolesov, G. V. Leplyanin, *Russ. J. Appl. Chem.*, 76 [4], *582-584*, 2003

[Salom1987] C. Salom, A. Horta, I. Hernández-Fuentes, I. F. Piérola, *Macromolecules*, 20, *696-698*, 1987

[Sarkar2008] S. Sarkar, A. Chunder, W. Fei, L. An, L. Zhai, *J. Am. Ceram. Soc.*, 91 [8], *2751–2755*, 2008

[Scheffler2002] M. Scheffler, Q. Wei, E. Pippel, J. Woltersdorf, P. Greil, *Key Eng. Mat.*, 206-213, *289-292*, 2002

[Scheffler2003] M. Scheffler, E. Pippel, J. Woltersdorf, P. Greil, *Mater. Chem. Phys.*, 80, *565–572*, 2003

[Schneider1988] O. Schneider, *Thermochim. Acta*, 134, *269-274*, 1988

[Schulz2004] M. Schulz, M. Börner, J. Göttert, T. Hanemann, J. Haußelt, G. Motz, *Adv. Eng. Mater.*, 6, 8, *676-680*, 2004

[Seok1998] W. K. Seok, L. G. Sneddon, *Bull. Korean. Chem. Society*, 19 [12], 1998

[Shah2002] S. R. Shah, R. Raj, *Acta Mater.*, 50, *4093–4103*, 2002

[Shah2005] S.R. Shah, R. Raj, *J. Eur. Ceram. Soc.*, 25, *243–249*, 2005

[Shinohara1998] H. Shinohara, Y. Yamakita, K. Ohno, *J. Mol. Struct.*, 442, *221-234*, 1998

[Stantschev2005] G. Stantschev, M. Frieß, R. Kochendörfer, W. Krenkel, *J. Eur. Ceram. Soc.*, 25, *205–209*, 2005

[Sun2002] Y.-P. Sun, B. Zhou, K. Henbest, K. Fu, W. Huang, Y. Lin, S. Taylor, D. L. Carroll, *Chem. Phys. Lett.*, 351, *349-353*, 2002

[Sun2007] Sun X, Zhang J, Zhang X, Lu S, Wang X, *J. Lumin.*, 955, *122–123*, 2007

[Suzuki1997] M. Suzuki, Y. Nakata, H. Nagai, T. Okutani, N. Kushibiki, M. Murakami, *Mater. Sci. Eng. B*, 49, *172-174*, 1997

[Suzuki1998] Suzuki, H.; Hoshino, S.; Yuan, C.-H.; Fujiki, M.; Toyoda, S.; Matsumoto, N. *Thin Solid Films*, 331, *64-70*, 1998

[Suzuki2005] T. Suzuki, G. Senthil Murugan, Y. Ohishi, *Appl. Phys. Lett.*, 86, *131903*, 2005

[Takahashi2001] T. Takahashi, J. Kaschta, H. Münstedt, *Rheol. Acta*, 40, *490-498*, 2001

[Takahashi2003] T. Takahashi, P. Colombo, *J. Porous Mat.*, 10, *113–121*, 2003
[Thünemann2007] M. Thünemann, O. Beffort, S. Kleiner, U. Vogt, *Compos. Sci. Technol.*, 67, *2377–2383*, 2007
[Trassl2002] S. Trassl, G. Motz, E. Rössler, G. Ziegler, *J. Am. Ceram. Soc.*, 85 [1], *239*, 2002
[Trimmel2003] G. Trimmel, R. Badheka, F. Babonneau, J. Latournerie, P. Dempsey, D. Bahloul-Houlier, J. Parmentier, G. D. Soraru, *J. Sol-Gel Sci. Techn.*, 26, *279–283*, 2003
[Vakifahmetoglu2009] C. Vakifahmetoglu, I. Menapace, A. Hirsch, L. Biasetto, R. Hauser, R. Riedel, P. Colombo, *Ceram. Int.*, 35, *3281-3290*, 2009
[Valenta2004] J. Valenta, N. Lalic, J. Linnros, *Appl. Phys. Lett.*, 84 [9], *1459-1461*, 2004
[Vaucher2008]S. Vaucher, J. Kuebler, O. Beffort, L. Biasetto, F. Zordan, P. Colombo, *Compos. Sci. Technol.*, 68, *3202–3207*, 2008
[Vij2006] D. R. Vij, *Applied Solid State Spectroscopy*, Springer, 2006
[Wacker] MK Technical data sheet, last accessed 08.10.09
[Wan2001] J. Wan, M. J. Gasch, A. K. Mukherjee, *J. Am. Ceram. Soc.*, 84 [10], *2165–2169*, 2001
[Wan2002] J. Wan, M. J. Gasch, A. K. Mukherjee, *J. Am. Ceram. Soc.*, 85 [3], *554–564*, 2002
[Wan2005] J. Wan, A. Alizadeh, S. T. Taylor, P. R. L. Malenfant, M. Manoharan, S. M. Loureiro, *Chem. Mater.*, 17 [23], *5613-5617*, 2005
[Wan2006] J. Wan, R.-G. Duan, M. J. Gasch, A. K. Mukherjee, *Mater. Sci. Eng. A*, 424, *105–116*, 2006
[Wang2004] F. Wang, G. Dukovic, L. E. Brus, T. F. Heinz, *Phys. Rev. Lett.*, 92 [17], 2004
[Wang2005] W. Wang, Y. Lin, Y-P. Sun, *Polymer*, 46, *8634-8640*, 2005
[Wang2007] Z.G. Wang, X.T. Zu and L.M. Fang, *J. Nanopart. Res.*, 9, *289–292*, 2007
[Wang2008] Y. Wang, L. Zhang, W. Xu, T. Jiang, Y. Fan, D. Jiang, L. An, *J. Am. Ceram. Soc.*, 91 [12], *3971–3975*, 2008
[Wei2002] Q. Wei, E. Pippel, J. Woltersdorf, M. Scheffler, P. Greil, *Mater. Chem. Phys.*, 73, *281–289*, 2002
[Wilden2007] J. Wilden, G. Fischer, *Appl. Surf. Sci.*, 254, *1067–1072*, 2007
[Wilhelm2005] M. Wilhelm, C. Soltmann, D. Koch, G. Grathwohl, *J. Eur. Ceram. Soc.*, 25, *271–276*, 2005
[Yang2001] P. Yang, C. F. Song, M. K. Lü, J. Cjang, Y. Z. Wang, Z. X. Yang, G. J. Zhou, Z. P. Ai, D. Xu, D. L. Yuan, *J. Solid State Chem.*, 160, *272-277*, 2001
[Yang2004a] W. Yang, Z. Xie, J. Li, H. Miao, L. Zhang, L. An, *Solid State Commun.*, 132, *263–268*, 2004
[Yang2004b] W. Yang, H. Miao, Z. Xie, L. Zhang, L. An, *Chem. Phys. Lett.*, 383, *441–444*, 2004
[Yang2005] W. Yang, Z. Xie, H. Miao, , L. Zhang, H. Ji, L. An, *J. Am. Ceram. Soc.*, 88 [2], *466–469*, 2005
[Yang2006] W. Yang, Z. Fan, H. Wang, Z. Xie, H. Miao, L. An, *J. Ceram. Process. Res.*, 7 [4], *307-310*, 2006
[Yang2007] W. Yang, H. Wang, S. Liu, Z. Xie, L. An, *J. Phys. Chem. B*, 111, *4156-4160*, 2007
[Yang2008a] W. Yang, F. Gao, H. Wang, Z. Xie, L. An, *Cryst. Growth Des.*, 8 [8], *2606-2608*, 2008

[Yang2008b] W. Yang, F. Gao, H. Wang, X. Zheng, Z. Xie, L. An, *J. Am. Ceram. Soc.,* 91 [4], *1312–1315,* 2008

[Yang2008c] W. Yang, X. Cheng, H. Wang, Z. Xie, F. Xing, L. An, *Cryst. Growth Des.*, 8 [11], *3921-3923*, 2008

[Youngblood2002] G.E. Youngblood, D. J. Senor, R.H. Jones, S. Graham, *Compos. Sci. Technol.*, 62, *1127–1139*, 2002

[Zemanova2002] M. Zemanova, E. Lecomte, P. Säjgalík, R. Riedel, *J. Eur. Ceram. Soc.*, 22, *2963–2968*, 2002

[Zhu2007] C. Zhu, Yang Y, Liang X, Yuan S, Chen G, *J. Lumin.*, 126 [2], *707*, 2007

7. Conclusion

In the present thesis, a brand new class of photoluminescent materials was investigated. The work originated from the industrial need for new phosphors with innovating properties with respect to commercial ones. Homogeneous materials, moldable at low temperatures and thermally stable up to 150 °C are required. In this study we could demonstrate that the heat-treated silicon-based materials are suitable candidates, as they provide all these properties. They are homogeneous, thermally stable and can be crosslinked at low temperatures. However, the photoluminescent properties of heat-treated silicon-based polymers have not been investigated before. Four classes of Si-polymers were analyzed in this thesis, which are generally used as precursors for polymer derived ceramics (PDCs): polysiloxanes, polysilazanes, polysilylcarbodiimides and a copolymer composed of a polysilazane and a polysilylcarbodiimide. Intrinsic fluorescence properties were detected in the materials. The different precursors investigated presented several affinities after heat-treatment, indicating that the degradation process that converts them to the pyrolyzed state is similar for all systems. The photoluminescence properties were investigated after heat-treatment at different temperatures, as well as the structural properties, via several characterization methods. In the polymers that do not contain aromatic substituents, such as the polysiloxane MK and the polysilazanes Ceraset and VL20, two main structural changes were correlated with the development of photoluminescence properties: crosslinking of the polymers and formation of sp^2 carbon. The presence of dangling bonds was initially considered a likely origin of the luminescence, as it occurs for similar materials, for example sol-gel glasses. EPR measurements demonstrated that carbon radicals are present in heat-treated silicon-based polymers but comparison with photoluminescence spectra indicated that they do not have an appreciable influence on the development of the fluorescence. Therefore, the discussion was focused on the relationship between the photoluminescence properties and the amount of crosslinking and free carbon, respectively. Although the effect of crosslinking on photoluminescence was demonstrated, the explanation of the nature of the luminescent centers formed by means of crosslinking is still an open issue. On the contrary, the influence of the formation of sp^2 carbon, in the form of small aromatic agglomerations, on the photoluminescence and absorption spectra could be clarified. The fluorescence

7. Conclusion

properties of the annealed precursors containing various amounts of sp^2 carbon were compared with the known emission of carbon nanotubes in different concentrations and analogies were ascertained.

An interesting polysilazane from the luminescence point of view is the KiON S, which shows visible emission obtained after annealing at a temperature as low as 300 °C, but the aromatic solvent, which contained the KiON S in the commercial package, is still present after drying and after annealing, and influences the fluorescence properties. Therefore, the polymer was not considered in the photoluminescence analysis.

The further two systems analyzed, namely the polysilylcarbodiimides (S1, S2, S3 and S4) and the copolymer GV, obtained from the reaction of VL20 and S1, are characterized by the presence of phenyl groups as substituents on the silicon atoms, and its known emission was monitored after heat-treatment. By comparison of the photoluminescence spectra after annealing of the four polysilylcarbodiimides, the effect of crosslinking on the emission ranges was attested. Furthermore, also in this system the role of small aromatic agglomerations was confirmed. The annealed copolymer GV demonstrated photoluminescence improvements with respect to its individual precursors, which open the possibility of further optimizations in order to obtain even higher quantum efficiencies. In this case, many luminescence centers are active at the same time. The emission bands due to the phenyl groups were monitored after heat-treatment and the effect of crosslinking on the phenyl groups emission was confirmed. Moreover, a new efficient luminescence center develops by means of crosslinking. As in the other systems, the effect of free carbon is detectable at higher annealing temperatures.

Nevertheless, it was demonstrated that most of the samples reveal a change in the fluorescence emission after storage in air for 1-2 years. Relatively stable fluorescence emission was detected mainly in the samples heat-treated from 500-600 °C. In view of the application in LEDs, the fluorescence emission should not change with time: application in protective atmosphere is also an option. On the contrary, features such as thermal stability, mouldability and homogeneity were proven to characterize the heat-treated polymers.

This study is relevant also for the analysis of the polymer derived ceramics, as it deeply investigates the processes involved in the thermal transformation of the polymer to the ceramic state, i.e. crosslinking and formation of free carbon, in a range of treatment temperatures not extensively investigated so far.

In summary, the present thesis contains a fundamental work on the innovative class of photoluminescent heat-treated silicon-based polymers and demonstrates the achievement of thermally stable photoluminescent polymers showing efficient visible emission after heat-treatment at low temperatures, suitable for LED applications.

8. Outlook

In the present thesis, several photoluminescent heat-treated silicon-based polymers were investigated and common features were highlighted. Some photoluminescence mechanisms were discerned and correlated to structural changes on the molecular scale. Due to our recent discovery, further detailed investigations are required in order to complete the photoluminescence analysis of the heat-treated silicon-based polymers.

The effect of free carbon formation on photoluminescence properties was elucidated in this work. However, the structure of the freshly formed free carbon and its development processes has to be further analyzed. The development of new fluorescence centers by means of crosslinking reactions was revealed; however, their origin is still unclear and needs further investigations.

Moreover, the improvement of the quantum efficiency is essential for the industrial application of the novel materials. The clarification of the new luminescence center formed by means of crosslinking of the GV polymer could suggest how to proceed to obtain high emission intensities.

One of the goals of the thesis was to produce materials insensitive to air. Only the samples heat-treated at higher temperatures turned out to be insensitive in this respect. Nevertheless, use in protective atmospheres could be the solution for the application of these materials.

Polymers with fully known structure should be investigated in the future. Therefore, model compounds should be synthesized. In this way, the fluorescence centers formed could be better understood and studied. Functional groups that provide crosslinking, for example vinyl groups, and photoluminescence properties, i.e. phenyl groups, should be grafted to the polymer chain. First steps performed in this direction were presented in chapters 5.1.2. and 5.1.3. Polysiloxanes were chosen because of their air stability and commercial availability. The goal is the production of air and thermally stable, moldable and efficiently fluorescent materials for applications in LEDs.

The thermal stability of the heat-treated samples and of their fluorescence properties has to be accurately investigated. Long term thermal analysis or a long term annealing should be performed on the heat-treated samples at the maximum temperature reached by the LED device (150 °C), in order to simulate the working conditions of the LED. Additionally, the working atmosphere can vary. In order to have a good thermal stability for LED applications, the fluorescence properties of the samples should not be influenced by the treatment.

Heat-treated samples in film and bulk form could also be produced and their fluorescence properties measured. Furthermore, other silicon-based polymers could be investigated by means of photoluminescence measurements. Further similarities or differences could help to analyse the photoluminescence mechanisms. Moreover, the influence of the annealing atmosphere, temperature and substituents should be also systematically verified.

Die VDM Verlagsservicegesellschaft sucht für wissenschaftliche Verlage abgeschlossene und herausragende

Dissertationen, Habilitationen, Diplomarbeiten, Master Theses, Magisterarbeiten usw.

für die kostenlose Publikation als Fachbuch.

Sie verfügen über eine Arbeit, die hohen inhaltlichen und formalen Ansprüchen genügt, und haben Interesse an einer honorarvergüteten Publikation?

Dann senden Sie bitte erste Informationen über sich und Ihre Arbeit per Email an *info@vdm-vsg.de*.

Sie erhalten kurzfristig unser Feedback!

VDM Verlagsservicegesellschaft mbH
Dudweiler Landstr. 99   Telefon +49 681 3720 174
D - 66123 Saarbrücken   Fax   +49 681 3720 1749
www.vdm-vsg.de

Die VDM Verlagsservicegesellschaft mbH vertritt

Printed by Books on Demand GmbH, Norderstedt / Germany